More Good Questions

Great Ways to Differentiate
Secondary Mathematics Instruction

More Good Questions

Great Ways to Differentiate
Secondary Mathematics Instruction

MARIAN SMALL *and* AMY LIN

Teachers College, Columbia University
New York and London

NATIONAL COUNCIL OF
TEACHERS OF MATHEMATICS
1906 Association Drive, Reston, VA 20191
www.nctm.org

www.nelson.com

Published simultaneously by Teachers College Press, 1234 Amsterdam Avenue, New York, NY 10027, and National Council of Teachers of Mathematics, 1906 Association Drive, Reston, VA 20191; distributed in Canada by Nelson Education Ltd., 1120 Birchmount Road, Toronto, ON, Canada M1K 5G4.

Text Design: Lynne Frost

Library of Congress Cataloging-in-Publication Data

Small, Marian.
    More good questions : great ways to differentiate secondary mathematics instruction / Marian Small and Amy Lin.
      p.   cm.
    Includes bibliographical references and index.
    ISBN 978-0-8077-5088-9 (pbk. : alk. paper)
      1. Mathematics—Study and teaching (Secondary).   2. Individualized instruction. 3. Effective teaching. I. Lin, Amy. II. Title.

    QA20.I53S634   2010
    510.71'2—dc22
                                        2010002447

ISBN 978-0-8077-5088-9 (paper)
NCTM Stock Number: 13782

Printed on acid-free paper
Manufactured in the United States of America

17 16 15 14 13 12 11   8 7 6 5 4 3

Contents

Preface

THIS BOOK EVOLVED from a similar volume at the prekindergarten through grade 8 level published last year (Small, 2009b). That resource was built on a long-term research study that looked at how students in the elementary years develop in their mathematical understanding. The research resulted, in part, in the creation of developmental maps that show the stages students move through in each of the strands of mathematics articulated by the National Council of Teachers of Mathematics (Small, 2005a, 2005b, 2006, 2007).

Having those maps led to the obvious question: What are they good for?

Clearly, a map of student development is most valuable if it can help teachers adjust instruction to meet the needs of students in their classes who are at different levels of development. With that in mind, we have created a number of professional development courses and many presentations wherein we shared with teachers two strategies that are relatively easy to implement and that allow them to make such curriculum adjustments. These strategies are open questions and parallel tasks.

Many teachers who have participated in these professional development sessions have reported that they found the approaches very useful and very manageable in the classroom environment, whether at the elementary or the secondary level. With their encouragement, and at their suggestion, we have written this book to provide even more models for how the strategies can be used.

ORGANIZATION OF THE BOOK

An introductory chapter describes the rationale for differentiating math instruction and explains the two principal strategies that are employed throughout the book: open questions and parallel tasks. Five content chapters then illustrate application of these strategies, followed by a final summary chapter, an appendix containing a template for teachers wishing to develop their own materials, a glossary, a bibliography, and an index.

Chapters 2–6 focus on the five content strands enunciated in the *Principles and Standards for School Mathematics* of the National Council of Teachers of Mathematics (NCTM):

- Algebra
- Number and Operations
- Geometry
- Measurement
- Data Analysis and Probability (NCTM, 2000)

The content strands are not developed sequentially, and Chapters 2–6 can be approached in any order.

Mathematical concepts are addressed in a framework of big ideas, which have been developed by marrying the NCTM process standards of problem solving, reasoning and proof, communicating, connecting, and representing (NCTM, 2000) with the NCTM content standards for the five content strands listed above. Big ideas are statements of fundamental principles and are broadly applicable to multiple grade bands and different developmental levels of understanding.

Within each of the content chapters, the suggested differentiating questions and tasks are divided according to middle and secondary grade bands set out in the NCTM (2000) standards:

- Grades 6–8
- Grades 9–12

The Appendix features a template worksheet that will assist teachers in developing their own materials in support of differentiated instruction through use of open questions and parallel tasks. An example of application of the worksheet appears in Chapter 1. The Glossary defines technical terms used throughout. Each word that appears in the Glossary is shown in boldface type at its first occurrence in the text, and each Glossary entry is annotated with the chapter and page number of the term's first occurrence.

The Bibliography highlights three types of resources: those that are referenced in other parts of the text, those that speak to the issues of teaching to big ideas and differentiating instruction, as well as a number of excellent sources for activities that can be used as is or used as a starting point for creating open questions and parallel tasks.

The Index focuses on educational concepts—standards, student development, teaching methods and principles, and such—as opposed to mathematical concepts. To facilitate user access to the mathematical topics covered, an Index of Big Ideas is provided, listing all big ideas covered in the content chapters.

A number of topics will be met in several strands. Using the Index and Glossary, together, will help the reader find where questions related to these topics can be found. For example, there are questions about parallelism in both the Measurement and the Geometry chapters, questions about factoring in both the Number and Operations and the Algebra chapters, and questions about similarity in both the Measurement and the Geometry chapters.

ORGANIZATION OF THE CONTENT CHAPTERS

Chapters 2–6 address the five NCTM content strands, providing examples of open questions and parallel tasks—organized around big ideas—for the grades 6–8 and 9–12 grade bands.

Each chapter begins with a listing of the goals of the NCTM standard for the particular content strand, followed by a brief description of how student understanding of the content develops across grade levels from grades 6 through 12. For each grade band, concepts students will learn and apply are described, demonstrating how basic concepts form the foundation for more complex concepts as understanding develops. The content standards are approached through exploration of big ideas, which are listed at the beginning of each chapter.

The bulk of each chapter is composed of a series of illustrations of application of the two differentiation strategies that are the focus of this book: open questions and parallel tasks. Each of these strategies is discussed for each of the two grade bands. Within each grade band section, content is organized by big idea, often with several illustrations for each big idea. Readers may choose to focus on the grade band in which they teach, or they may choose to study both of the grade bands to get a better sense of how understanding of the content develops in students.

For many of the questions and tasks described, important follow-up questions that can be used in the classroom are listed. In many cases, variations of activities or follow-up questions are provided to allow for even more flexibility in differentiating instruction. In addition, the rationale for each activity is presented as background that will make it easier for teachers to develop their own related activities.

Numerous Teaching Tips are included in each chapter. These sometimes relate to a specific activity, but often they are general strategies that can apply to any situation in which a teacher is attempting to differentiate instruction in math.

At the end of each chapter, concluding remarks include a few suggestions for developing additional open questions and parallel tasks by using the template provided in the Appendix.

AS THE AUTHORS, it is our hope that teachers will embrace the two core strategies—open questions and parallel tasks—that are explained and demonstrated in this book, and find them, as we have, to be helpful to the many children who come into classrooms with highly differentiated mathematical preparation, skill, and confidence. Seeing a child who has been struggling in mathematics start to feel successful is an important goal in our teaching. We have seen the use of the strategies described in this volume make that happen over and over again.

Acknowledgments

I HAVE BEEN PRIVILEGED to have had a number of educators across the United States and Canada respond very positively to a book on differentiating instruction through good questions for prekindergarten through grade 8 that I authored last year. They have been very encouraging about the need for a similar book for teachers at higher grade levels. That encouragement led to this book.

In particular, I thank a group of educators in Ontario, Canada, who have been instrumental in pushing forward the approaches that are the focus of this book, for the grades 7–12 teachers in that province. I have been fortunate to work closely with educators like Myrna Ingalls, in the Ministry of Education, and a number of other consultants, such as Amy Lin—my coauthor for this volume—who have become wonderful friends and colleagues. Because of that work, I have had a chance to try out many of these ideas with students and teachers and can attest to their success.

This book is also a direct result of the opportunity provided to me by Jean Ward, an editor at Teachers College Press who has invited me to work on this book. I thank her for her confidence in me and her assistance in the preparation of this book.

—*Marian Small*

I HAVE BEEN VERY FORTUNATE to have met and have had the opportunity to work with Dr. Marian Small. Our friendship and professional collaboration through different projects have meant a great deal to me, and Marian has truly brought out the creative ideas from me for this book.

I am grateful to have worked closely with two math coaches, Ruth Teszeri and Todd Malarczuk, who were the inspiration for my work on questioning. The job-embedded professional learning they carried out with teachers helped me to see firsthand the success of powerful and purposeful questions used in the mathematics classroom.

I am also thankful that my husband, Andrew, and my two sons, Jeremy and Zachary, who were always supportive and patient in spite of the time this work took me away from them.

—*Amy Lin*

Why and How to Differentiate
Math Instruction

STUDENTS IN ANY CLASSROOM differ in many ways, only some of which the teacher can reasonably attend to in developing instructional plans. Some differences will be cognitive—for example, what previous concepts and skills students can call upon. Some differences will be more about learning style and preferences, e.g., whether the student learns better through auditory, visual, or kinesthetic approaches. Other differences will be more about preferences, including behaviors such as persistence or inquisitiveness or the lack thereof and personal interests.

THE CHALLENGE IN MATH CLASSROOMS

Although teachers in other subject areas sometimes allow students to work on alternative projects, it is much less likely that teachers vary the material they ask their students to work with in mathematics. The math teacher will more frequently teach all students based on a fairly narrow curriculum goal presented in a textbook. The teacher will recognize that some students need additional help and will provide as much support as possible to those students while the other students are working independently. Perhaps this occurs because **differentiating instruction** in mathematics is a relatively new idea. Perhaps it is because teachers may never have been trained to really understand how students differ mathematically. However, students in the same math classroom clearly do differ mathematically in significant ways. Teachers want to be successful in their instruction of all students. Understanding differences and differentiating instruction are important processes for achievement of that goal.

The National Council of Teachers of Mathematics (NCTM), the professional organization whose mission it is to promote, articulate, and support the best possible teaching and learning in mathematics, recognizes the need for differentiation. The first principle of the NCTM *Principles and Standards for School Mathematics* reads, "Excellence in mathematics education requires equity—high expectations and strong support for all students" (NCTM, 2000, p. 12).

In particular, NCTM recognizes the need for accommodating differences among students, taking into account both their readiness and their level of mathematical talent/interest/confidence, to ensure that each student can learn important mathematics. "Equity does not mean that every student should receive identical instruction; instead, it demands that reasonable and appropriate accommodations be made as needed to promote access and attainment for all students" (NCTM, 2000, p. 12).

THE PARTICULAR CHALLENGE IN GRADES 6–12

The challenge for teachers of grades 6–12 is even greater than in the earlier grades, particularly in situations where students are not streamed. Although there is much evidence of the value, particularly for the struggling student, of being in heterogeneous classrooms, the teacher in those rooms must deal with significant student differences in mathematical level. While some students are still struggling with their multiplication facts or addition and subtraction with decimals, others are comfortable with complex reasoning and problem solving involving fractions, decimals, and percents. The differences between students' mathematical levels, beginning as far back as kindergarten or grade 1, continue to be an issue teachers must face all through the grades.

Where some see the answer as streaming, many believe that the answer is a differentiated instruction environment in a destreamed classroom.

WHAT IT MEANS TO MEET STUDENT NEEDS

One approach to meeting each student's needs is to provide tasks within each student's **zone of proximal development** and to ensure that each student in the class has the opportunity to make a meaningful contribution to the class community of learners. Zone of proximal development is a term used to describe the "distance between the actual development level as determined by independent problem solving and the level of potential development as determined through problem solving under adult guidance or in collaboration with more capable peers" (Vygotsky, 1978, p. 86).

Instruction within the zone of proximal development allows students, whether with guidance from the teacher or by working with other students, to access new ideas that are beyond what the students know but within their reach. Teachers are not using educational time wisely if they either are teaching beyond the student's zone of proximal development or are providing instruction on material the student already can handle independently. Although other students in the classroom may be progressing, the student operating outside his or her zone of proximal development is often not benefiting from the instruction.

For example, a teacher might be planning a lesson on calculating the whole when a percent that is greater than 100% of the whole is known, using a problem such as asking students to determine what number 30 is 210% of. Although the skill that the teacher might emphasize is solving a proportion such as

$$\frac{210}{100} = \frac{30}{x}$$

the more fundamental objective is getting students to recognize that solving a percent problem is always about determining a ratio equivalent to one where the second term is 100.

Although the planned lesson is likely to depend on the facts that students can work algebraically with two fractions, one involving a variable, and that they understand the concept of a percent greater than 100%, a teacher could effectively teach a meaningful lesson on what percent is all about even to students who do not have those abilities. The teacher could allow the less developed students to explore the idea of determining equivalent ratios to solve problems using percents less than 100% with ratio tables or other more informal strategies (rather than formal proportions) while the more advanced students are using percents greater than 100% and more formal methods. Only when the teacher felt that the use of percents greater than 100% and algebraic techniques were in an individual student's zone of proximal development would the teacher ask that student to work with those sorts of values and strategies. Thus, by making this adjustment, the teacher differentiates the task to locate it within each student's zone of proximal development.

ASSESSING STUDENTS' NEEDS

For a teacher to teach to a student's zone of proximal development, first the teacher must determine what that zone is. This can be accomplished by using prior assessment information in conjunction with a teacher's own analysis to ascertain a student's mathematical developmental level. For example, to determine an 8th-grade student's developmental level in working with percents, a teacher might use a diagnostic to find out whether the student interprets percents as ratios with a second term of 100, relates percents to equivalent fractions and/or decimals, can represent a percent up to 100% visually, can explain what 150% means, and recognizes that solving a percent problem involves determining an equivalent ratio.

Some tools to accomplish this sort of evaluation are tied to developmental continua that have been established to describe students' mathematical growth (Small, 2005a, 2005b, 2006, 2007, 2010). Teachers might also use locally or personally developed tools to learn about students' prior knowledge. Only after a teacher has determined a student's level of mathematical sophistication, can he or she meaningfully address that student's needs.

PRINCIPLES AND APPROACHES TO DIFFERENTIATING INSTRUCTION

Differentiating instruction is not a new idea, but the issue has been gaining an ever higher profile for mathematics teachers in recent years. More and more, educational systems and parents are expecting the teacher to be aware of what each individual student—whether a struggling student, an average student, or a gifted student—needs and to plan instruction to take those needs into account. In the past, this was less the case in mathematics than in other subject areas, but now the expectation is common in mathematics as well.

There is general agreement that to effectively differentiate instruction, the following elements are needed:

- **Big Ideas.** The focus of instruction must be on the **big ideas** being taught to ensure that they are addressed, no matter at what level (Small, 2009a; Small & Lin, 2010).
- **Prior Assessment.** Prior assessment is essential to determine what needs different students have (Gregory & Chapman, 2007; Murray & Jorgensen, 2007).
- **Choice.** There must be some aspect of choice for the student, whether in content, process, or product.

Teaching to Big Ideas

The Curriculum Principle of the NCTM *Principles and Standards for School Mathematics* states that "A curriculum is more than a collection of activities: it must be coherent, focused on important mathematics, and well articulated across the grades" (NCTM, 2000, p. 14).

Curriculum coherence requires a focus on interconnections, or big ideas. Big ideas represent fundamental principles; they are the ideas that link the specifics. For example, the notion that benchmark numbers are a way to make sense of other numbers is equally useful for the 6th-grader who is trying to place −22 on a number line, the 8th-grader who relates π to the number 3.14, or the 10th-grader who is trying to estimate the sine of a 50° angle. If students in a classroom differ in their readiness, it is usually in terms of the specifics and not the big ideas. Although some students in a classroom where estimating the value of radicals is being taught may not be ready for that precise topic, they could still deal with the concept of estimating and why it is useful in simpler situations.

Big ideas can form a framework for thinking about "important mathematics" and supporting standards-driven instruction. Big ideas cut across grade bands. There may be differences in the complexity of their application, but the big ideas remain constant. Many teachers believe that curriculum requirements limit them to fairly narrow learning goals and feel that they must focus instruction on meeting those specific student outcomes. Differentiation requires a different approach, one that is facilitated by teaching to the big ideas. It is impossible to differentiate too narrow an idea, but it is always possible to differentiate instruction focused on a bigger idea.

Prior Assessment

To determine the instructional direction, a teacher needs to know how students in the classroom vary in their mathematical developmental level. This requires collecting data either formally or informally to determine what abilities or what deficiencies students have. Although many teachers feel they lack the time or tools to undertake prior assessment on a regular basis, the data derived from prior assessment should drive how instruction is differentiated.

Despite the importance of prior assessment, employing a highly structured approach or a standardized tool for conducting the assessment is not mandatory. Depending on the topic, a teacher might use a combination of written and oral questions and tasks to determine an appropriate starting point for each student or to determine what next steps the student requires. An example of one situation is described below, along with steps teachers might take, given the student responses.

Consider the task below:

> The Beep-Beep pager company charges $30 to set up a client with a pager along with a $7.50 monthly fee. The Don't Miss Them pager company charges $9 a month, but no set up fee. How long would you have to own a pager before the Beep-Beep deal is a better one?

Students might respond to the task in very different ways. Here are some examples:

- Joshua immediately raises his hand and just waits for the teacher to help him.
- Blossom says that $30 + 7.50 + 9 = 46.50$, so it would take 46.50 months.
- Madison writes: $y = 30 + 7.50x$ and $y = 9x$, so $9x = 30 + 7.5x$. That means $1.5x = 30$, and $x = 30 \div 1.5$. Since $x = 20$, it would be 20 months.
- Lamont starts a table like this one, but forgets to add the terms to answer the question.

Beep-Beep	Don't Miss Them
37.50	9
7.50	9
7.50	9
7.50	9

- Hannah uses an appropriate table, and extends it until the Beep-Beep value is less and counts the number of entries.

Beep-Beep	Don't Miss Them
37.50	9
45	18
52.50	27
60	36
. . .	. . .
180	180
187.50	189

- Latoya reasons that the difference in price is $1.50 a month, so you just divide 30 by 1.50 to figure out how many months it would take to make up the extra cost at the start. She calculates the value to be 20 and then indicates that after 21 months (assuming whole numbers of months), the Beep-Beep plan is better.

The teacher needs to respond differently based on what has been learned about the students. For example, the teacher might wish to:

- Encourage Joshua to be more independent or set out an alternate related problem that is more suitable to his developmental level
- Help Blossom understand that just because there are three numbers in a problem, you don't automatically add, and emphasize the importance of reading carefully what the problem requests
- Encourage Madison's thoughtful approach to the problem, but help her see that she still hasn't really answered the question posed
- Ask Lamont to label his columns and tell what each represents, then ask him how the table might help him solve the problem
- Ask Hannah to think of a way she could have used her idea without having to show every single row in the table
- Ask Latoya, who clearly is thinking in a very sophisticated way, to create a different scenario where the Beep-Beep plan would not be better until, for example, 31 months

By knowing where the students are in their cognitive and mathematical development, the teacher is better able to get a feel for what groups of students might need in the way of learning and can set up a situation that challenges each individual while still encouraging each one to take risks and responsibility for learning (Karp & Howell, 2004).

Choice

Few math teachers are comfortable with the notion of student choice except in the rarest of circumstances. They worry that students will not make "appropriate" choices.

However, some teachers who are uncomfortable differentiating instruction in terms of the main lesson goal are willing to provide some choice in follow-up activities students use to practice the ideas they have been taught. Some of the strategies that have been suggested for differentiating practice include the use of menus from which students choose from an array of tasks, tiered lessons in which teachers teach to the whole group and vary the follow-up for different students, learning stations where different students attempt different tasks, or other approaches that allow for student choice, usually in pursuit of the same basic overall lesson goal (Tomlinson, 1999; Westphal, 2007).

For example, a teacher might present a lesson on using the exponent laws to simplify expressions to all students, and then vary the follow-up. Some students might work only with simple situations; these tasks are likely to involve simple multiplications of pairs of numbers with the same base, such as $2^5 \times 2^7$. Other students might be asked to work with situations where a variety of laws might be called on at once, such as simplifying $2^5 \times (2^7)^2 \div 2^5$. Some students might deal with even more challenging questions, such as determining two factors for 1 million where neither one involves a power of 10 (e.g., $10^6 = 2^6 \times 5^6$). By using prior assessment data, the teacher is in a better position to provide appropriate choices.

TWO CORE STRATEGIES FOR DIFFERENTIATING MATHEMATICS INSTRUCTION: OPEN QUESTIONS AND PARALLEL TASKS

It is not realistic for a teacher to try to create 30 different instructional paths for 30 students, or even 6 different paths for 6 groups of students. Because this is the perceived alternative to one-size-fits-all teaching, instruction in mathematics is often not differentiated. To differentiate instruction efficiently, teachers need manageable strategies that meet the needs of most of their students at the same time. Through the use of just two core strategies, teachers can effectively differentiate instruction to suit all students. These two core strategies are the central feature of this book:

- **Open questions**
- **Parallel tasks**

Open Questions

The ultimate goal of differentiation is to meet the needs of the varied students in a classroom during instruction. This becomes manageable if the teacher can create a single question or task that is inclusive not only in allowing for different students to approach it by using different processes or strategies but also in allowing for students at different stages of mathematical development to benefit and grow from attention to the task. In other words, the task is in the appropriate zone of proximal development for the entire class. In this way, each student becomes part of the larger learning conversation, an important and valued member of the learning community. Struggling students are less likely to be the passive learners they so often are (Lovin, Kyger, & Allsopp, 2004).

A question is open when it is framed in such a way that a variety of responses or approaches are possible. Consider, for example, these two questions, each of which might be asked of a whole class, and think about how the responses to each question would differ:

> *Question 1:* Write the quadratic $y = 3x^2 - 12x + 17$ in vertex form.
>
> *Question 2:* Draw a graph of $y = 3x^2 - 12x + 17$. Tell what you notice.

Question 1 is a fairly closed question. If the student does not know what vertex form is, there is no chance he or she will answer Question 1 correctly. In the case of Question 2, a much more open question, students simply create the graph and notice whatever it is that they happen to notice—whether that is the vertex, that the shape is parabolic, that it opens upward, and so on.

Strategies for Creating Open Questions. This book illustrates a variety of styles of open questions. Some common strategies that can be used to construct open questions are described below:

- Turning around a question
- Asking for similarities and differences
- Replacing a number, shape, measurement unit, and so forth with a blank
- Asking for a number sentence

Turning Around a Question. For the turn-around strategy, instead of giving the question, the teacher gives the answer and asks for the question. For example:

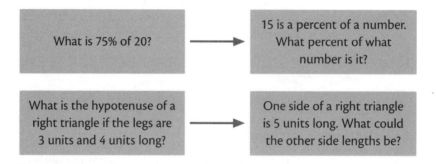

Asking for Similarities and Differences. The teacher chooses two items—two numbers, two shapes, two graphs, two probabilities, two measurements, and so forth—and asks students how they are alike and how they are different. Inevitably, there will be many good answers. For example, the teacher could ask how the number $\sqrt{2}$ is like the number $\sqrt{5}$ and how it is different. A student might realize that both are irrational numbers, both are less than 3, both are greater than 1, and both are side lengths of squares with a whole number of units of area.

Replacing a Number with a Blank. Open questions can be created by replacing a number (or numbers) with a blank and allowing the students to choose the number(s) to use. For example, instead of asking for the surface area of a cone with radius 4" and height 15", the teacher could ask students to choose numbers for the radius and height and then determine the surface area. By allowing choice, the question clearly can go in many directions. Most importantly, students can choose values in such a way that their ability to demonstrate understanding of the concept being learned is not compromised by extraneous factors such as the complexity of the calculations required of them.

Asking for a Sentence. Students can be asked to create a sentence that includes certain words and numbers. For example, a teacher could ask students to create a sentence that includes the number 0.5 along with the words "sine," "rational," and "amplitude," or a sentence that includes the words "linear" and "increasing" as well as the numbers 4 and 9. The variety of sentences students come up with will often surprise teachers. For example, for the second situation, a student might produce any of the sentences below and many more:

- An *increasing* *linear* pattern could include the numbers 4 and 9.
- In a *linear* pattern starting at 4 and *increasing* by 9, the tenth number will be 85.
- A *linear* pattern that is *increasing* by 9 grows faster than one that is *increasing* by 4.

Shortcut for Creating Open Questions. A teacher can sometimes create an open question by beginning with a question already available, such as a question from a text resource. Here are a few examples:

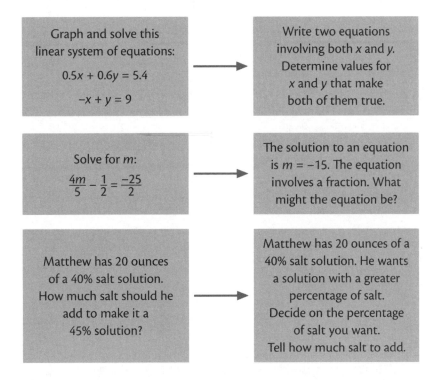

What to Avoid in an Open Question. An open question should be mathematically meaningful. There is nothing wrong with an occasional question such as *What do you like about algebra?* but questions that are focused more directly on big ideas or on curricular goals are likely to accomplish more in terms of helping students progress satisfactorily in math.

Open questions need just the right amount of ambiguity. They may seem vague, and that may initially bother students, but the vagueness is critical to ensuring that the question is broad enough to meet the needs of all students.

On the other hand, teachers must be careful about making questions so vague that they deter thinking. Compare, for example, a question like *What is infinity?* with a question like *How do you know that there are an infinite number of decimals between 0 and 1?* In the first case, a student does not know whether what is desired is a definition for the word, something philosophical, or the symbol ∞. The student will most likely be uncomfortable proceeding without further direction. In the second case, there is still ambiguity. Some students may wonder if a particular approach is desired, but many students will be comfortable proceeding by using their own strategies.

The reason for a little ambiguity is to allow for the differentiation that is the goal in the use of open questions. Any question that is too specific may target a narrow level of understanding and not allow students who are not at that level to engage with the question and experience success.

*A **Different Kind of Classroom Conversation.*** Not only will the mathematical conversation be richer when open questions are used, but almost any student will be able to find something appropriate to contribute.

The important point to notice is that the teacher can put the same question to the entire class, but the question is designed to allow for differentiation of response based on each student's understanding. All students can participate fully and gain from the discussion in the classroom learning community.

This approach differs, in an important way, from asking a question, observing students who do not understand, and then asking a simpler question to which they can respond. By using the open question, students gain confidence; they can answer the teacher's question right from the start. Psychologically, this is a much more positive situation.

*A **Multiple Benefits.*** There is another benefit to open questions. Many students and many adults view mathematics as a difficult, unwelcoming subject because they see it as black and white. Unlike subjects where students are asked their opinions or where they might be encouraged to express different points of views, math is viewed as a subject where either you get it or you don't. This view of mathematics inhibits many students from even trying. Once they falter, they lose confidence and assume they will continue to falter—they may simply shut down.

It is the responsibility of teachers to help students see that mathematics is multifaceted. Any mathematical concept can be considered from a variety of perspectives, and those multiple perspectives actually enrich its study. Open questions provide the opportunity to demonstrate this.

*A **Fostering Effective Follow-Up Discussion.*** Follow-up discussions play a significant role in cementing learning and building confidence in students. Thus, it is important for teachers to employ strategies that will optimize the effectiveness of follow-up discussions to benefit students at all developmental levels.

To build success for all students, it is important to make sure that those who are more likely to have simple answers are called on first. By doing so, the teacher will increase the chances that these students' answers have not been "used up" by the time they are called on.

The teacher must convey the message that a variety of answers are appreciated. It is obvious to students when a teacher is "looking for" a particular answer. An open question is designed to ensure that many answers are good answers and will be equally valued.

The teacher should try to build connections between answers that students provide. For example, when asked how sine and cosine are alike, one student might say that both functions only have values between –1 and 1 and another might say that they are both trigonometric ratios. The teacher could follow up with:

* *What makes them ratios? What are the two terms?*
* *How does looking at the terms of the ratio tell you the values are between –1 and 1?*
* *When are the values negative?*

Such questions challenge all students and scaffold students who need help.

Parallel Tasks

Parallel tasks are sets of tasks, usually two or three, that are designed to meet the needs of students at different developmental levels, but that get at the same big idea and are close enough in context that they can be discussed simultaneously. In other words, if a teacher asks the class a question, it is pertinent to each student, no matter which task that student completed. The use of parallel tasks is an extension of Forman's (2003) point that task modification can lead to valuable discussions about the underlying mathematics of a situation. Parallel tasks also contribute to the creation of the classroom as a learning community in which all students are able to contribute to discussion of the topic being studied (Murray & Jorgensen, 2007).

For example, suppose a teacher wishes to elicit the big idea within the NCTM Geometry strand that it is useful to apply transformations to analyze mathematical situations. The teacher can set out two parallel tasks:

> *Option 1:* Draw a triangle ABC in Quadrant II of a coordinate grid. Reflect it so that the image is in Quadrant IV. Describe your reflection line.
>
> *Option 2:* Draw a triangle ABC in Quadrant II of a coordinate grid. Reflect it so that the image is in Quadrant IV. Determine the matrix that describes the transformation.

Both options focus on the concept of using transformations to analyze mathematical situations, but *Option 1* is suitable for students not yet comfortable with matrices. Further, the tasks fit well together because questions such as the ones listed below suit a discussion of both tasks and thus can be asked of all students in the class:

- *How did you know that what you had performed was a reflection?*
- *Did the line of reflection have a positive or negative slope? Why?*
- *Why can you describe the image of any point using just one piece of information in addition to the point's coordinates (either the line or the matrix)?*

Strategies for Creating Parallel Tasks. To create parallel tasks to address a particular big idea, it is important to make use of prior assessment data to determine how students might differ developmentally in approaching that idea. Differences might relate to what operations the students can use or what size numbers they can handle, or they might involve, for example, what meanings of an operation make sense to the students.

Once the developmental differences have been identified, the object is to develop similar enough contexts for the various options that common questions can be asked of the students as they reflect on their work. For example, for the big idea that using a smaller unit requires more of those units, the major developmental difference might be the type of measurement conversions with which students are comfortable. One task could focus on linear measurements and another on area measurements. One set of parallel tasks might be:

> **_Option 1:_**   Someone suggests that the school driveway is 4,000,000 mm long. Is it a long driveway?
>
> **_Option 2:_**   Someone suggests that a shopping mall might be 4,000,000 cm² in area. Do you think that's reasonable?

In this case, common follow-up questions could be:

- *Is it easy to imagine how big your measurement actually is?*
- *Why does it help to think of it in terms of other units?*
- *What other units did you choose? How did you rewrite the measurement in those units?*
- *How can you tell whether your answer is reasonable?*

Often, to create a set of parallel tasks, a teacher can select a task from a handy resource (e.g., a student text) and then figure out how to alter it to make it suitable for a different developmental level. Then both tasks are offered simultaneously as options for students:

> **Original task** (e.g., from a text):
>   486 students voted in the school election. That was about 53% of the student body. How many students are in the school?
>
> **Parallel task:**
>   486 students voted in the school election. That was about 60% of the student body. How many students are in the school?

Common follow-up questions could be:

- *How do you know that there are more than 500 students?*
- *How do you know that there are fewer than 1,000 students?*
- *Why might someone do division to answer the question? What division?*
- *How might you estimate your answer?*
- *What does the question have to do with looking for equivalent ratios or fractions?*
- *How many students did you decide were in the school? How did you decide?*

Fostering Effective Follow-Up Discussion. The role of follow-up discussions of parallel tasks and the techniques for encouraging them mirror those for open questions. Once again, it is critical that the teacher demonstrate to students that he or she values the tasks equally by setting them up so that common questions suit each of them. It is important to make sure students realize that the teacher is equally interested in responses from both groups of students. The teacher should try not to call first on students who have completed one of the tasks and then on students who have completed the other(s). Each question should be addressed to the whole group. If students choose to identify the task they selected, they may, but it is better if the teacher does not ask which task was performed as the students begin to talk.

Management Issues in Choice Situations. Some teachers are concerned that if tasks are provided at two levels, students might select the "wrong" task. It may indeed be appropriate at times to suggest which task a student might complete. This could be done by simply assigning a particular task to each student. However, it is important sometimes—even most of the time—to allow the students to choose. Choice is very empowering.

If students who struggle with a concept happen to select a task beyond their ability, they will soon realize it and try the other option. Knowing that they have the choice of task should alleviate any frustration students might feel if they struggle initially. However, students may also sometimes be able to complete a task more challenging than they first thought they could handle. This would be a very positive experience.

If students repeatedly select an easier task than they are capable of, they should simply be allowed to complete the selected task. Then, when they are done, the teacher can encourage them privately to try the other option as well.

Putting Theory into Practice

A form such as the ones that appear on the next two pages can serve as a convenient template for creation of customized materials to support differentiation of instruction in math. In this example, a teacher has developed a plan for differentiated instruction on the topic of algebra. A blank form is provided in the Appendix.

MY OWN QUESTIONS AND TASKS

Lesson Goal: Solving a system of linear equations   **Grade Level:** __9__

Standard(s) Addressed:

Understand and apply concepts and procedures of algebra.

Underlying Big Idea(s):

Many equivalent representations can describe the same pattern or generalization. Each representation may give more insight into certain characteristics of the situation or generalization.

Open Question(s):

The solution to a system of linear equations is $x = 5$ and $y = 2$.
What might the equations be?

Parallel Tasks:

Option 1:

Solve for x and y:

$$2x + y = 17$$
$$2x - y = 15$$

Option 2:

Solve for x and y:

$$2.5x - 3.5y = -0.75$$
$$-4x + 1.7y = -22.2$$

Principles to Keep in Mind:

- All open questions must allow for correct responses at a variety of levels.
- Parallel tasks need to be created with variations that allow struggling students to be successful and proficient students to be challenged.
- Questions and tasks should be constructed in such a way that will allow all students to participate together in follow-up discussions.

MY OWN QUESTIONS AND TASKS

Lesson Goal: Determine side lengths of triangular   **Grade Level:** __8__
spaces using scale drawings

Standard(s) Addressed:

Use ratios and proportions to represent real-world situations.

Underlying Big Idea(s):

Knowing the measurement of one shape can sometimes provide
information about the measurements of another.

Open Question(s):

A right triangle is a scale drawing of a triangular piece of land.

The scale drawing has one side length of 12".

The piece of land has one side length of 18'.

What could the other side lengths of the triangular piece of land be?

Parallel Tasks:

The right triangle on the left
is a scale drawing of a
triangular piece of land.   **Option 1:**

What is length x?

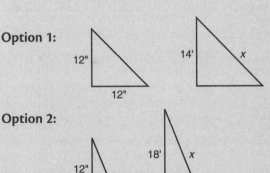

Option 2:

Principles to Keep in Mind:

• All open questions must allow for correct responses at a variety of levels.

• Parallel tasks need to be created with variations that allow struggling students
to be successful and proficient students to be challenged.

• Questions and tasks should be constructed in such a way that will allow all
students to participate together in follow-up discussions.

The following fundamental principles should be kept in mind when developing new questions and tasks:

- All open questions must allow for correct responses at a variety of levels.
- Parallel tasks need to be created with variations that allow struggling students to be successful and proficient students to be challenged.
- Questions and tasks should be constructed in such a way that all students can participate together in follow-up discussions.

Teachers may find it challenging at first to incorporate the core strategies of open questions and parallel tasks into their teaching routines. However, after trying examples found in the five chapters that follow, and creating their own questions and tasks, teachers will soon find that these strategies become second nature. And the payoff for the effort will be the very positive effects of differentiation that emerge: fuller participation by all students and greater advancement in learning for all.

CREATING A MATH TALK COMMUNITY

Throughout this book, many suggestions will be offered for ways to differentiate instruction. These are all predicated on a classroom climate where mathematical conversation is the norm, a variety of student approaches are encouraged and valued, and students feel free to take risks. Unless a student engages in mathematical conversation, it is not possible for a teacher to know what that individual does or does not understand. Consequently, it is not possible for a teacher to know what steps must be taken to ensure the student's needs are being met.

Some teachers may be nervous about offering choices or asking open questions—worried that students might get off track, worried that students might be uncomfortable with not being absolutely certain about what is expected, or worried that students may offer an idea with which the teacher is unfamiliar, leaving the teacher unsure about how to proceed. These are natural concerns.

Initially, students who are accustomed to highly structured learning environments may find open questions or choice unsettling. But once the students see the teacher's willingness to allow them to go in different directions and see that the teacher honors all reasonable contributions, whether "correct" or not, they will grow comfortable with the change and will appreciate the opportunity for greater input. Teachers will also find it both surprising and rewarding to see how students rise to the challenge of engaging in mathematical conversation and how students often help the teacher sort out an unclear comment from another student or suggest ways to pick up on another student's suggestion.

Algebra

DIFFERENTIATED LEARNING activities in algebra are derived from applying the NCTM process standards of problem solving, reasoning and proof, communicating, connecting, and representing to content goals of the NCTM Algebra Standard, including

- understanding patterns, **relations**, and **functions**
- representing and analyzing mathematical situations and structures using algebraic symbols
- using mathematical models to represent and understand quantitative relationships
- analyzing change in various contexts (NCTM, 2000)

TOPICS

Before beginning the task of differentiating student learning in algebra, it is useful for teachers to have a good sense of how the topics in the strand develop over the grade bands. The NCTM *Curriculum Focal Points* (NCTM, 2006), which suggest what mathematical content should be the focus at each grade level through grade 8, were used as the basis for recommendations made in this resource for grades 6–8; the NCTM *Principles and Standards for School Mathematics* (NCTM, 2000) helped form the basis for the material for the higher grades. For a teacher at a particular grade level, it can be helpful to be aware of where students' learning is situated in relation to what learning has preceded the present grade band and what will follow.

Grades 6–8

Within this grade band, students intensify their work with traditional algebra. They use **expressions**, **equations**, and formulas to correspond to numerical and real-life situations. They evaluate expressions involving **variables** and use variables more regularly, recognizing that two different expressions might be equivalent.

They solve simple equations and use **tables of values** to uncover relationships.

Grades 9–12

Within this grade band, students study classes of relationships or functions, particularly **linear**, **quadratic**, **polynomial**, **rational**, **exponential**, and **trigonometric functions**. They learn what the elements of each class have in common, how they are related, and in what situations they are useful.

They develop a much deeper **symbol sense** and solve a much broader range of equations using many techniques, including algebraic and graphical ones.

THE BIG IDEAS FOR ALGEBRA

Coherent curricula in algebra that meet NCTM content and process standards (NCTM, 2000) and support differentiated instruction can be structured around the following big ideas:

- Algebraic reasoning is a process of describing and analyzing generalized mathematical relationships and change using words and symbols.
- Comparing mathematical patterns or relationships either algebraically or graphically helps us see that there are classes of relationships with common characteristics and helps us describe each member of the class.
- The same **algebraic expression** or pattern can be related to different situations, and different algebraic expressions can describe the same real-world situation, sometimes with limitations based on context.
- Many equivalent representations can describe the same pattern or generalization. Each representation may give more insight into certain characteristics of the situation or generalization.
- Limited information about a mathematical pattern or relationship can sometimes, but not always, allow us to predict other information about that pattern or relationship.
- The principles and processes that underlie operations with numbers and solving equations involving numbers apply equally to algebraic situations.
- The **transformations** that are fundamental to determining the geometric relationships between shapes apply equally to algebraic situations.

The tasks set out and the questions asked while teaching algebra should be developed to evoke these ideas. The following sections present numerous examples of application of open questions and parallel tasks in development of differentiated instruction in these big ideas across two grade bands.

CHAPTER CONTENTS			
Open Questions: Grades 6–8	19	Parallel Tasks: Grades 6–8	46
Open Questions: Grades 9–12	29	Parallel Tasks: Grades 9–12	52

OPEN QUESTIONS FOR GRADES 6–8

> **OPEN QUESTIONS** are broad-based questions that invite meaningful responses from students at many developmental levels.

✹ **BIG IDEA.** Algebraic reasoning is a process of describing and analyzing generalized mathematical relationships and change using words and symbols.

> A number pattern includes both a 3 and a 13 as terms. What might be the **general term** of the pattern?

Students have a great deal of flexibility in approaching this problem. First, they can think about the characteristics of the pattern, because this would influence the general term. For example, they can decide that 3 and 13 are the first two numbers in the pattern, that the pattern is the set of all odd whole numbers $(1, 3, 5, 7, \ldots)$, that the pattern is simply the set of counting numbers $(1, 2, 3, \ldots)$, or something much different—for example, that the pattern consists of alternating positive and negative **integers** $(1, -2, 3, -4, 5, -6, \ldots)$.

To create the general term, students will have to think algebraically. For example, for the odd numbers, they recognize that each number in the pattern is 1 less than the corresponding numbers in $2, 4, 6, 8, \ldots$. The numbers in the even number pattern are just doubles, of the form $2n$, so the general term for the odd number pattern must be $2n - 1$.

If students do not know how to begin, the teacher could ask scaffolding questions such as:

- *Where did you decide to put the 3? As the first term or somewhere else in the pattern?*
- *Does the 3 have to come before the 13?*
- *What does general term mean?*
- *What picture might you draw or what comparison might you make to come up with the general term?*

Variations. An alternative way to prompt students to find the general term could take the form of a question such as the following: *A pattern includes a figure made of 3 squares and a figure made of 13 squares. How many squares might the 100th figure have? How do you know?* Rather than working all the way to the 100th figure "by hand," students will quickly see the virtue of an algebraic approach.

> An algebraic expression is described using words. Some of the words in the pattern are:
>
> *opposite, less, 4, shared*
>
> What might the expression be?

In this situation, students are connecting algebraic expressions to their verbal translations. A student should realize that the word "shared" probably suggests a division, that the word "opposite" probably connotes using a negative sign somewhere, and that the word "less" might involve subtraction, but not necessarily. For example, expressions and translations such as the following might be offered:

$\dfrac{-3x-2}{4}$ *2 less than the opposite of triple a number is shared among 4 people*

$\dfrac{-4-x}{8}$ *The opposite of 4 less a number is shared among 8*

> A **pattern rule** includes the following words and numbers (among others), not necessarily in this order:
>
> *2, subtract, multiply*
>
> Write out a complete pattern rule.
>
> List the first 10 terms in your pattern based on your rule.
>
> Tell about a number greater than 100 that is NOT in your pattern and how you know it's not there.

This question is open because there is so much latitude in what the pattern rule might be. It is also open because students can choose the number greater than 100 that is not in the pattern. For example, a student might write the pattern rule: *Multiply the term number by 4 and then subtract 2.* Noticing that the numbers are all even, the student might state that 101 is not possible in the pattern. Alternately, a student might say the rule is: *Multiply the term number by 2 and subtract 1.* The student might then indicate that 200 is never in the pattern because 200 + 1 is not twice anything.

Variations. It is easy to vary this question by requiring other words or numbers to be part of the pattern rule. These might include *add, divide, 15,* and so on.

✹ **BIG IDEA.** Comparing mathematical patterns or relationships either algebraically or graphically helps us see that there are classes of patterns or relationships with common characteristics and helps us describe each member of the class.

Which two of these patterns are most alike? Why?

> Pattern 1:   2, 5, 8, 11, 14, . . .
>
> Pattern 2:   2, 7, 12, 17, 22, 27, . . .
>
> Pattern 3:   3, 6, 9, 12, 15, . . .

Some students will decide that Patterns 1 and 2 are most alike because they both start with 2 (i.e., they belong to the class of patterns where the first term is 2). Others will decide that Patterns 1 and 3 are most alike because they both go up by 3s. It might be an interesting challenge for students to decide if there is some way to argue that Patterns 2 and 3 are most alike; perhaps they will notice that those patterns both include the number 27.

Variations. This question could be varied by using other triples or even groups of four patterns. Or, instead of numerical patterns, figure patterns could be presented, such as the ones shown below.

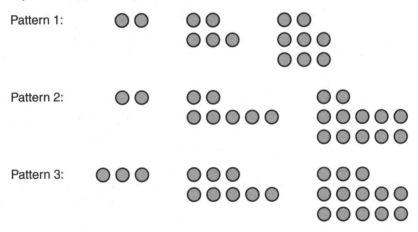

Jamie lists the first five terms of a linear pattern that grows quickly, and Adrienne lists the first five terms of a linear pattern that grows slowly.

What could their patterns be? How could the graphs representing the patterns differ?

Students could be provided with tiles to help them create visual patterns if this would be useful for some. Students who use tiles will probably create patterns involving whole numbers, where the terms in the pattern represent the number of tiles in each figure. Students who do not use visual representations might create problems with whole numbers, decimals, integers, fractions, and so forth. All students will have to remember or be reminded of what linear patterns are.

The variety of patterns students might create is vast, and there is a great deal of latitude in deciding what quick growth versus slow growth looks like. For example, a student might say that $1, 2, 3, 4, 5, \ldots$ grows slowly compared with $100, 200, 300, 400, 500, \ldots$. Another student might suggest that $1, 2, 3, 4, 5, \ldots$ grows quickly compared with $0.001, 0.002, 0.003, 0.004, 0.005, \ldots$.

A rich discussion about the comparative nature of the terms quick and slow is likely to ensue. Students could be asked how they know that the patterns selected are linear and to justify their characterization of relative growth rates. Students might observe that the difference in rates can be hard to detect graphically if different scales are used for their graphs; with similar scales, the differences are quite easy to observe.

Variations. Students might also be asked to examine the algebraic expressions that represent their patterns and discuss how those representations differ. They might also consider whether or how the slow growth or quick growth could have been predicted from the algebraic representations.

> How are these two equations alike? How are they different?
>
> $$y = 3x - 2 \qquad y = 6x - 4$$

This question is very open in that students might simply observe that the equations are alike because both involve subtraction and are different because different numbers are involved as **coefficients** and **constants**. With some encouragement to use their understanding that equations describe relationships, some students might notice that for a given x value the resulting y in the second equation is always double the y in the first equation.

An exploration of this question could also lead to consideration of the fact that all terms of both sides of an equation must be multiplied by the same value to maintain the same relationship. In other words, the equations would have had the same solution only if the second equation had been $2y = 6x - 4$.

> What makes all of the algebraic expressions below similar?
>
> $$5x \qquad 4x + 5 \qquad x^2 \qquad 30 - x$$

This question requires students to consider how the represented relationships behave. Some students might notice that one thing they have in common is that when $x = 5$, the expressions are all worth 25. But some students will notice other things, for example, that they all involve the variable x (which allows an easy entry point for students), that they include both numbers and letters, or that if you **substitute** 1 for *x*, the results are all odd numbers.

For students who struggle to come up with a response, questions that could be asked include:

- *If the value of* x *is small, are the values of the expression big or small?*
- *What values could you substitute for* x?
- *Do you ever get the same values in different expressions when you do a substitution?*

TEACHING TIP. Providing a set of items and asking how they are similar is often an easy way to create an open question. It is valuable, however, if there is more than one way to "sort" the items.

✴ **BIG IDEA.** The same algebraic expression or pattern can be related to different situations, and different algebraic expressions can describe the same real-world situation, sometimes with limitations based on context.

> You describe a situation with the expression 5x. What might the situation be?

It is important that students see how algebraic expressions or equations model real-world situations. This particular expression is fairly simple. One possible scenario might be the value of a given number of $5 bills; another might be the number of fingers on a given number of hands; another might be the cost of five movie tickets if each ticket costs $*x*.

If students have difficulty getting started, the teacher could provide scaffolding with questions such as these:

- *How could you describe the expression in words?*
- *What operation is hidden inside the expression?*
- *When would you use that operation?*

Variations. Instead of a simple expression like 5*x*, one might use slightly more complex expressions such as $2x + 30$, or equations such as $5x = 40$ or $2x + 30 = 40$.

> A rectangle has a length of *x*" and a width of 4". What algebraic expressions can describe features of the rectangle?

Relating expressions to measurement formulas is often a good connection to algebra for students. They recognize that they have been doing algebra when using formulas and just not realizing it.

In this situation, a student might use the expression $2x + 8$ to describe the **perimeter** of the rectangle, the expression $4x$ to describe the area, the expression $x - 4$ to tell how much longer the rectangle is than it is wide, or the expression $4x + 8$ to describe the perimeter if two of the rectangles were attached along a 4" side.

TEACHING TIP. Connecting algebra with other content strands is very helpful for students who seem to struggle with the abstractness of the algebra strand.

✹ **BIG IDEA.** Many equivalent representations can describe the same pattern or generalization. Each representation may give more insight into certain characteristics of the situation or generalization.

> A pattern starts like this:
>
> 4, 7, 10, 13, . . .
>
> How could you represent the pattern with a picture to help someone see why the next numbers might be 16 and 19?

It is important that students see that certain representations of patterns and relationships might provide more immediate insight into what is happening than other representations.

For example, a representation like the one below makes it clear not only that 16 and 19 are next, but that any term value is 1 greater than 3 times the term number. This is because the first row gives the first number, the first two rows together give the second number, the first three rows together give the third number, and so on.

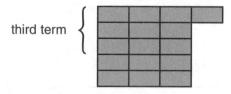

third term {

Other representations are possible, of course, but it is the responsibility of the student to discuss why his or her representation is useful for seeing why the next terms might be 16 and 19.

For students who do not know how to begin, questions such as these could be helpful:

- *How could you use colored cubes to show the first number? The second number?*
- *How could you use color to emphasize how the pattern is changing?*
- *Does your picture make it easy to predict the next term?*

You want to use a picture to show why the equations $3x + 2 = 7$ and $3x + 3 = 8$ really say the same thing. What picture would you draw? How does that picture help show why they say the same thing?

This question focuses students on how certain representations of relationships provide insight into those relationships. A student might draw a balance scale with 3 bags and 2 counters on one side balancing 7 counters on the other side, and then add another counter to each side. Alternatively, a student might draw a picture like the one below. Notice that the rows ($3x + 3$ and 8) are of equal length, and that they would also be of equal length if the 1s at the end of each row were dropped.

x	x	x	2	1
7				1

The question is open in that students can choose any representation they wish.

Variations. The question could be made even more open if students are allowed to choose two equations that they think are equivalent and then draw a picture to demonstrate that equivalence.

TEACHING TIP. Allowing students some choice in which equations they represent provides easier access for some students than suggesting what equation must be represented.

✸ **BIG IDEA.** Limited information about a mathematical pattern or relationship can sometimes, but not always, allow us to predict other information about that pattern or relationship.

The 13th term of a **linear growing pattern** is at least 30 more than the 5th term. What do you know about the 15th term of the pattern?

This question is open in that students might provide a variety of answers. One student might, for example, simply say that the 15th term is more than the 13th one. Another student might suggest that the 15th term is more than 7 more than the 13th term, figuring that if you go up by more than 30 going from the 5th term to the 13th term, then you go up more than one fourth as much when going only two terms further. Either answer shows some understanding of what a growing pattern is.

The fact that the difference between the 5th and 13th terms is given as "at least 30" rather than using a specific number makes the question even more open.

Variations. The difference between terms could be given as "about 30" or "less than 30" instead, or the difference might be between other terms, not the 5th and 13th.

TEACHING TIP. Using a phrase like "at least 30" instead of a specific number makes a question more open than a question using a specific number. It is also useful to ask what students know rather than asking for a specific piece of information.

> The 8th term of a linear pattern has a value of 20. What could the algebraic expression for the general term be?

The pattern created is the choice of the student; the only constraint is that the 8th term is 20 and that term values increase or decrease by a fixed amount. Some students will simply work down from 20 to get a first term, using a chosen increase or decrease; others may first determine a relationship between 8 and 20. For example, a student who notices that $20 = 2 \times 8 + 4$ might simply use a pattern rule of $2n + 4$.

If students struggle to come up with a response, the teacher could ask scaffolding questions such as:

- *What makes a pattern linear?*
- *Could the pattern decrease or must it increase?*
- *Could the first term be 0? If so, what would the pattern be?*

Variations. Instead of requiring a linear pattern, different types of patterns could be allowed. The value of the term and the position of the term could also be altered. In another variation, the question could be posed as: *A linear growing pattern is created with figures made of squares. The 8th figure uses 20 squares. How many squares might the nth term have?*

> You know that using a certain rule, the number 8 turns into the number 40 and the number 12 turns into the number 52.
>
> What do you think the number 10 might turn into? Why?
>
> What about the number 20?

Because no absolute rule is given, there are many appropriate responses to this situation. Some students will simply say that they think 10 turns into 46, since 10 is halfway between 8 and 12 and 46 is halfway between 40 and 52. They might say that 20 is likely to turn into 92, since 20 is twice 10 and 92 is twice 46. This is quite reasonable.

Other students will notice that 40 is 16 more than 3×8, just like 52 is 16 more than 3×12, and suggest that 10 turns into 46, since $46 = 3 \times 10 + 16$, and 20 turns into 76, since $76 = 3 \times 20 + 16$.

For students having difficulty getting started, questions that could be asked include:

- *What relationship is there between 8 and 40?*
- *Does that same relationship hold for 12 and 52?*
- *What other relationship is there between 8 and 40?*
- *Did you think of, perhaps, combining two operations?*

The fifth term of a pattern is a design that looks like this:

What might the first three terms look like? Why?

Describe the pattern.

There are many directions a student might go. For example, any of the choices below would make sense:

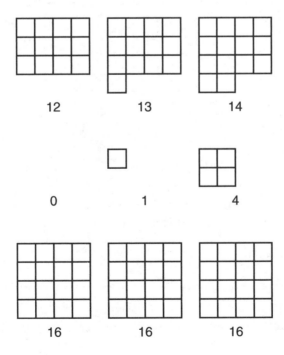

| 12 | 13 | 14 |

| 0 | 1 | 4 |

| 16 | 16 | 16 |

Students have opportunities to consider a variety of types of patterns, including patterns that grow by a constant amount, patterns that decrease by a constant amount, or even square number patterns.

If students do not know how to begin, the teacher could provide scaffolding with questions such as these:

- *Can you be sure whether the pattern increases or decreases or neither?*
- *Does the pattern have to change by the same amount each time?*
- *Could the first design in the pattern be made up of 20 squares?*

✹ BIG IDEA. **The principles and processes that underlie operations with numbers and solving equations involving numbers apply equally to algebraic situations.**

> Why does it make sense that $3x - 5x = -2x$?

When asked why $3x - 5x$ can be **simplified** to $-2x$, students will inevitably make a connection to what they know about whole numbers. But their explanations can certainly vary.

One student might think of using **algebra tiles** and point out that if you have $3x$, you can try to remove $5x$. This only works if a form of zero, particularly $-2x + 2x$ is added. The student is recognizing that subtraction still means take away, whether using algebraic expressions or numbers.

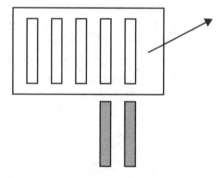

Another student might use the notion that if you have $3x$, it is $2x$ fewer than if you have $5x$, so the answer is $-2x$. That student, too, is using what he or she knows about operations with numbers.

> The **product** of the length and width of a rectangle is 4*xy*. What might the length and width be?

Students who realize how numbers are multiplied and how, for example, 4·5·6, or 120, can be written as 4·30 or 5·24 or 6·20, might realize that they can multiply 4 by *xy* or 4*x* by *y* or 4*y* by *x*. Other students will break up the 4 and suggest a possibility is 2 multiplied by 2*xy* or maybe 2*x* by 2*y*.

Variations. Rather than asking for values that multiply to a particular product, one can ask for expressions that lead to a specific sum, difference, or **quotient** instead.

TEACHING TIP. Giving the answer and asking for the question is a "fail-safe" strategy for creating open questions.

OPEN QUESTIONS FOR GRADES 9–12

✻ BIG IDEA. **Algebraic reasoning is a process of describing and analyzing generalized mathematical relationships and change using words and symbols.**

> What do you notice about the graphs of lines of the form $y = mx + m$? Why does what you notice make sense?

Suggest that students graph a number of lines of the form $y = mx + m$ on their graphing calculators.

Opening up a question by asking if it makes sense engages all students in exploring the situation and providing reasons for their answer. In this question, students know that the **slope** and **y-intercept** are the same value. They can investigate the situation visually and then use either visual or algebraic techniques, for example, solving the **simultaneous linear equations** $y = mx + b$ and $b = m$, to talk about why what happens does happen (i.e., all of the lines intersect at $(-1,0)$). Graphs can be drawn by hand or using technology.

> One side of a **right triangle** is part of the line with equation $y = -2$. What could be the equations of the lines of which the other two sides are part?

Providing a horizontal side for the triangle makes the problem more accessible to struggling students. Students could, if they chose, use a vertical line as another side and then only have to do significant work to determine one equation. However, the problem could be as challenging as students want to make it.

> Start with rectangles with length *l* and width *w*. Increase the length by any amount you wish and decrease the width by the same amount. How does the area usually change?

By allowing students to choose values for the length, width, and increase/decrease and allowing them to choose how many rectangles to try, the question becomes more open to more students. During the course of the discussion, students will notice that if the greater value is *l*, then the area decreases each time. Eventually, the whole class might consider the expression $(l + x)(w - x)$, to see that it results in $lw - [x^2 + (l - w)x]$, helping to explain the decrease from *lw*. Students might come to this realization by using a visual such as the one shown at the right.

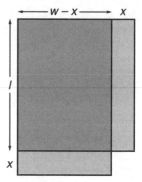

lw changes to
$lw - lx + xw - x^2$, or
$lw - [x^2 - (l - w)x]$

> 5, 7, 9, 11, 13, . . . is a **sequence**.
>
> Zack said that the formula is $t_1 = 5, t_n = t_{n-1} + 2$.
>
> Eric said that a better way to write the formula is $t_n = 2n + 3$.
>
> Do you agree or disagree with Eric? Why?

This type of open question lets students be correct in choosing either statement and then supporting the choice with their arguments. The discussion around who is right will emphasize the differences and similarities between **recursion formulas** and **explicit formulas** for a sequence.

✳ **BIG IDEA.** Comparing mathematical patterns or relationships either algebraically or graphically helps us see that there are classes of relationships with common characteristics and helps us describe each member of the class.

> Which of these four relationships are most alike? Why?
>
> $y = 4x + 5$ $y = 4x + 3$ $y = 3x + 5$ $y = 4x - 3$

This question is quite open because students could reasonably have different opinions. Some will suggest that the second and fourth are most alike because they involve $4x$ and 3 and all that changes is the sign. Others will argue that the first two are most alike because both involve adding something to $4x$. Others will argue that the first and third are most alike because they involve adding 5.

Once students graph all of these relations on their graphing calculators, they might find further support for their argument or might even change their minds.

> How is a quadratic relationship the same as a linear relationship? How are they different?

Rather than being specific and asking about particular comparisons such as **first differences**, this more open question asking students to compare and contrast the two relationships allows more possibilities for student engagement. Students might choose to discuss the graphs, the equations, or patterns relating to both quadratic and linear functions.

For students who struggle to come up with a response, questions such as these could be helpful:

* *What makes a relationship linear?*
* *What makes it quadratic?*
* *What do the graphs of these relationships typically look like?*
* *What real-life examples do you know that are described by linear or quadratic relationships?*
* *What patterns create linear relationships?*
* *How do these differ from patterns that create quadratic relationships?*

> Graph $y = x^2 + 4x + 4$ and $y = 4(x - 4)^2 - 1$. Tell everything you notice about both equations and both graphs.

Both equations use the variable x, involve squaring, and use the values 1, 4, and 4, so even students not fully comfortable with quadratics will notice some commonalities. The two equations represent different forms of the quadratic equation, a more **standard form** and a **vertex form**, allowing for a rich discussion of what features different representations of a quadratic make easy to access.

Graphs can be created using technology (e.g., graphing calculator, graphing software) or by hand. By asking students to tell everything they know about the graphs, the teacher will be able to accept a range of answers from descriptions about the shape of the graph, how wide or narrow the graphs are, how high or low they are, their **vertices**, their **axes of symmetry**, and so forth.

Variations. Students can be asked to graph $y = 2x^2 + 2x + 2$ and $y = 4x^2 + 4x + 4$. Then they can tell everything they notice about both graphs.

> Two **parabolas** have the same *x-intercepts* (−2,0) and (4,0). The maximum or minimum value of the first parabola is two times the maximum or minimum value of the other parabola. Sketch these parabolas on the same **coordinate grid**.

This problem gives students choice on where they place their parabolas given the constraints on the *x*-intercepts and the maximum and minimum values of each. Shapes of the parabola are not affected by the *x*-intercepts or where the maximum and minimum values end up. Sketching parabolas is fairly easy when given the *x*-intercepts.

During the discussion, students will notice that in each case, the vertex was at (1,*a*). They can be encouraged to discuss why this is so, thus focusing on the **symmetry** of parabolas.

> Compare the **roots** of these three equations. What do you notice?
>
> $4x^2 - 17x + 4 = 0$ $\qquad$ $6x^2 - 37x + 6 = 0$ $\qquad$ $8x^2 - 65x + 8 = 0$
>
> Add another equation that acts the same way.

This question asks students to describe what they observe as they investigate this special case for a quadratic equation where the roots are **reciprocals**. Most students will notice that the magnitude of the coefficient of *x* is 1 more than the product of the magnitudes of the coefficient of x^2 and the constant.

Students might solve the problem by **factoring**, using the **quadratic formula**, or graphing. They will notice that even within the class of quadratic functions, there are subclasses. Some might end up investigating other subclasses, for example, those where the roots are negative reciprocals.

If students have difficulty getting started, the teacher could ask scaffolding questions such as:

- *What do you notice about the evenness or oddness of the numerical values in the equations?*
- *Do you think there is a root near 0? Why or why not?*
- *Do you think there are any negative roots? Why or why not?*

> Is the graph of $y = 3x^2 + 4x - 5$ more like the graph of $y = 3x^2 - 5$ or more like the graph of $y = 4x - 5$? Explain.

Students might use graphing calculators to compare different sections of the graphs (e.g., sections near *x* = −1,000, near *x* = 0, and near *x* = 1,000) to see if they would draw the same conclusions for all sections of the graph.

This question allows students to compare the three graphs and draw their own conclusions about which are more alike. In the process, all students are likely to consider the difference between linear and quadratic relations.

> Draw a graph of a parabola that grows quickly and the graph of a parabola that grows slowly. What are the equations of these parabolas?

The question gets students thinking about the shape of a parabola if it grows more quickly, or when the absolute value of the coefficient of x^2 is greater than 1. Then they think about a parabola that grows more slowly, or when the absolute value of the coefficient of x^2 is less than 1. When they work on developing equations for these parabolas they will consider whether their parabolas are opening up or down, how "fat" or "skinny" they are, and where the vertices of each are placed. The question is open in that students can use their own judgment to decide what *quickly* and *slowly* mean. Students will also think about the fact that any parabola has one "side" that grows and one "side" that shrinks. They will have to discuss what they think *growth* means in this situation.

> You graph $y = 6x^2 + 5x + 1$.
>
> Does the graph change more if you increase the 6 by 1, the 5 by 1, or the 1 by 1?

An important concept for students to explore is the effect of the various **parameters** when working with classes of relations or functions. For example, when working with quadratics, students should be encouraged to understand the role of the *a*, the *b*, and the *c* in $y = ax^2 + bx + c$. One way to accomplish this is to ask a fairly open question, such as this one, which allows students to predict and then test their predictions.

Because this is an opinion question, there is no specific correct answer. This is a good example of a situation where graphing calculator technology is very useful. Students might draw very different conclusions about which graph changes the most when values of x are near 0 as compared to very large positive or negative values of x.

Variations. Students can investigate the comparative effects of altering the various parameters of other functions, whether trigonometric, rational, or polynomial.

✴ BIG IDEA. **The same algebraic expression or pattern can be related to different situations, and different algebraic expressions can describe the same real-world situation, sometimes with limitations based on context.**

> The tortoise and the hare start at the same place, and they each run at a constant speed. The tortoise wins the race. Use graphs or equations to describe the path of each racer.

This type of question allows choice in determining the distance traveled over time for the two racers. Students can choose how long the race is and how fast each animal is moving. Some students will use graphs or graphing technology to help them visualize the race. Others may create equations of lines, modifying the slope to show one racer moving more quickly than the other.

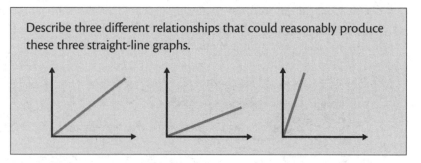

> Describe three different relationships that could reasonably produce these three straight-line graphs.

This question allows students to come up with their own stories to explain the differences between the three graphs, most likely focusing on the slopes of the lines. In each case, the y-intercept is 0, leading to a greater likelihood of success. Students might simply use three speeds of three different runners, three prices for three items, three linear growing patterns, and so forth.

Variations. Lines with negative slopes instead of positive slopes could be used, requiring quite different stories. Or, instead of using three lines, three parabolas or three exponential functions could be used.

TEACHING TIP. By removing the numbers and labels from a graph, a question becomes open to students and encourages them to think more conceptually. The strategy of not labeling graphs and **axes** can be used to open up many other tasks as well.

> Jeremy is crossing an intersection on his bike. How fast should the approaching car be going to avoid a collision?

This question allows students to create various scenarios in order to answer the question. Who is traveling faster? Where is the starting point for the bike? For the car? Graphing the situation will require consideration of both **partial** and **direct variations**. Students will decide on the conditions that allow for no collision to occur.

For students who do not know how to begin, questions that could be asked include:

- *Where is the starting point for the bike? For the car?*
- *Who is traveling faster?*
- *When will they meet?*
- *How can they avoid a collision?*
- *How might a graph help you decide this?*

> Jocelyn is starting her own dog walking business. She charges a fixed fee for any dog weighing 20 pounds or less. She charges an additional per-pound fee for dogs weighing more than 20 pounds. How does her profit change if she decides to charge more for either the fixed fee or the additional amount per pound?

This problem allows the student to choose the amount for the fixed fee and the additional amount charged per pound over the 20-pound limit. The question helps students understand the effects of changing each, or both, the slope and the *y*-intercept for a linear graph, depending on how they approach the problem. The question is open because students can choose their own fees for the business before investigating the effect of changing these rates on the overall profit.

TEACHING TIP. By not giving students values in a problem and asking them to choose the values, a question is opened up and students can start with numbers they are comfortable with.

> Keifer tossed a ball in the air, and after 3 seconds the ball reached its maximum height. Make a height (feet) versus time (seconds) graph. What is an equation that could model the path of the ball?

This question is an application of a quadratic relationship. A more traditional question of this type gives students the quadratic equation that describes the path of a projectile and then has them calculate the maximum height or the time it takes to reach the ground. In this question, students know how long it takes to reach the maximum height but can decide on where the ball is thrown from and the maximum height.

Variations. Keifer kicks a ball and clears a horizontal bar that is 3.3 yards above the ground. What is an equation that could model the path of the ball?

TEACHING TIP. Teachers often give two pieces of information in a problem and ask for a third. The simple change of offering only one piece of information and asking for two opens up the question.

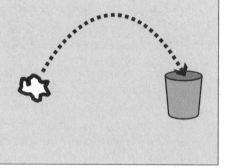

> A paper is crumpled into a ball and tossed into a wastebasket. Describe the path of the crumpled paper using a quadratic equation.
>
> How would you transform the equation so that the paper barely misses the basket?

A scenario is described but no dimensions or distances have been given. The parabolic path is shown and can be drawn on a graph to represent the quadratic function. Students have choices on the starting point (e.g., how high off the ground the paper is tossed from), the distance to the wastebasket, the velocity of the toss, the highest point of the paper's path, and how long the paper takes to get to the basket.

If students struggle to come up with a response, the teacher could provide scaffolding with questions such as these:

- *How do you know it's a parabola?*
- *Where is the person who is tossing the paper likely to be standing?*
- *If the path of the paper toss is traced on a graph, what would the axes be?*
- *How long does it take for the paper to be tossed into the bin?*
- *How does the force of the toss affect the trajectory of the paper?*

> The value of an investment, in thousands of dollars, is described with a rational function:
>
> $$f(t) = \frac{15t + 25}{t} \quad (t \text{ is in years})$$
>
> Tell what you know about this investment.

By asking the students what they know, the question is opened up for students to notice many aspects of the function that would tell something about the investment. Applying knowledge about the characteristics of the rational function and its graph to a real-world example not only helps students see where mathematics is actually used, but also allows students to start with what they are most familiar with. In the conversation with other students, each student will be likely to learn even more.

> Ruth took a wheel and marked a point on it with chalk. She rolled the wheel and recorded the height of the mark above the ground relative to the distance the wheel traveled. Sketch the resulting graph.

Students have the choice in deciding the size of their wheel and the starting point from where they will start measuring height versus the distance traveled. Some students will decide to have their starting point at the bottom of the wheel to make it easier to graph and others will choose the top or the sides. All choices of measurements will help students relate the **radius**, the height of the axle, and the distance traveled to characteristics of the graph.

For students having difficulty getting started, questions such as these could be helpful:

- *What is the radius of your wheel?*
- *Where did you place the mark on your wheel?*
- *What is the **amplitude** for the graph?*
- *What is the **period** for the graph? What does this represent?*

> Ferris wheel A is traveling faster than ferris wheel B. At the lowest point for one wheel a rider is 2 m above the ground, while the highest point for the other is 35 m above the ground.
>
> Sketch the height of a rider above the ground for two complete revolutions of both wheels on the same graph.

Typically, students working on problems like this one involving **sinusoidal functions** will be given the **diameter** of the wheel, the center above the ground, and the speed of the wheel and then will be asked to graph the function. This question is more open because of the choice in size and speed of the wheel. Some students may choose two ferris wheels that have the same dimensions and vary only the speed. Others may decide to choose different diameters and speeds. The resulting graph will compare the periods, amplitudes, and shapes of the graphs.

> The height of the rider in a large pendulum ride is given in the table.
>
> | Time (s) | 0 | 1 | 2 | 3 | 4 | 5 | 6 | 7 | 8 | 9 |
> |---|---|---|---|---|---|---|---|---|---|---|
> | Height (m) | 55 | 53 | 46 | 36 | 25 | 14 | 7 | 5 | 8 | 15 |
>
> Tell what you know about this pendulum ride.

When students are first introduced to sinusoidal functions, they must work to understand how the amplitude, the period, and **phase shifts** relate to the graph and equation for a particular function. This question begins with a real-life context, allowing students to begin with what they know to lead to discussions about aspects of sinusoidal functions with which they may be less familiar.

For this set of heights and times, students could identify amplitudes, vertical and phase shifts, and the period. Other students could determine the equation of the function, the rest positions of the pendulum, or the maximum displacement for the pendulum.

> There are close to 3,000 seats in a theater. There are 20 rows, and the numbers of seats in successive rows form an **arithmetic sequence**. How many seats can be in the front row?

To solve this problem, students need to create an **arithmetic series** that yields a sum that is close to 3,000 for $n = 20$. The question is open because students can create a number of different seating patterns that would yield an arithmetic series with 20 rows and a total number of seats near 3,000.

✳ **BIG IDEA.** Many equivalent representations can describe the same pattern or generalization. Each representation may give more insight into certain characteristics of the situation or generalization.

> When you model a certain algebraic expression with algebra tiles, it looks like a square. What could the expression be?

It is useful for students to recognize **perfect squares** even when they are in algebraic form. This question is open because one student can start with the tiles, make the squares, and record the algebraic expressions, while another could start with the algebra directly.

Students will undoubtedly notice that unless they use constants, for example, creating a square with 9 unit tiles, the resulting expression will have to be quadratic.

> Consider the equation $3y + 4x - 12 = 0$.
>
> What information does the graphical representation of the equation quickly provide that the algebraic representation does not provide as readily?

This question is simply one example of what a teacher might use to focus attention on the fact that different representations of a relationship can lead one to think differently about that relationship. Even when providing the same facts, one form might make it easier to see certain information than another form.

A student might suggest that the graph makes it much easier to see where the x-intercept and y-intercept are. Another might suggest that it makes it much easier to see why the y-value decreases when the x-value increases.

Or a student might suggest that the algebraic expression makes it much easier to see the exact y-coordinate that goes with a particular x-coordinate.

If students do not know how to begin, the teacher could ask scaffolding questions such as:

* *What does your graph look like?*
* *Was it obvious that you would get a line when you looked at the equation? What did you need to know to know that?*
* *Does y increase or decrease when x increases by 3? Why does that make sense? Is it easier to tell on the graph or with the algebraic expression?*

Variations. A graph might be provided instead of the algebraic expression. Then a student could consider what the algebraic expression makes it easier to see.

> These two equations are equivalent:
>
> $$2x + 5y = 400 \qquad x = -\frac{5}{2}y + \frac{400}{2}$$
>
> Liam says that both equations model this situation: *I have some $5 bills and two identical piles of $1 bills. Altogether, I have $400.*
>
> Which equation do you think is a better description? Why?

Some algebraic situations should, of course, involve contexts. Looking at this question might help students see that the equation used to represent a situation might be in a different form than the equation that makes calculations easiest.

> Using a total of 6 algebra tiles, represent a **trinomial**. Arrange the tiles into a rectangle and determine the factors of this trinomial.

The algebra tiles help students visualize the factors of a trinomial. This question opens up possible representations of any trinomial using a total of 6 tiles. All students can easily begin with the arrangement of the tiles into a rectangle to find the factors of the trinomial.

> What polynomials can you model using exactly 5 algebra tiles?

Students need to know that algebra tiles can be used to represent 1, x, and x^2 as well as their opposites.

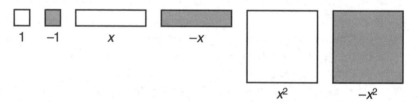

Then they can use different combinations of these tiles to model polynomials and, if they know the **zero principle** (i.e., that $x + (-x) = 0$), they can show more possibilities, such as $3x^2$ (represented as $3x^2 + x + (-x)$).

Variations. Students might be asked to represent polynomials with different numbers of tiles or to represent polynomials meeting certain conditions (e.g., using more negative terms than positive ones).

✴ **BIG IDEA.** Limited information about a mathematical pattern or relationship can sometimes, but not always, allow us to predict other information about that pattern or relationship.

> The slope of a line is $\frac{2}{3}$. Provide the **coordinates** for two points on the line.

Rather than giving two coordinates and asking for the slope, this question works backward. The question allows for the possibility of using any points on a line with this slope. Students can start with an equation with $m = \frac{2}{3}$ and read any two points they find on the line or they can start with one point, moving 3 over and 2 up, realizing that the equation of the line is unnecessary. Students can start with intercepts or even at (0,0).

For students who struggle to come up with a response, questions that could be asked include:

- *Does the line slant up or down? Is it steep?*
- *If you knew one point on the line, how would you move to get another point?*
- *Is there more than one line with this slope?*

Variations. Students could be asked the same question, with the further stipulation that the line must or must not pass through a specific point.

TEACHING TIP. Starting with the answer and asking the students for the question, as is done when giving a slope and asking for points, is a useful generic strategy for creating open questions.

A line passes through two of these points:

What could the equation
of the line be?

By not restricting the student to two particular points for determining the equation of the line, responses can vary according to which points the student finds easiest to use. Students can choose slopes to be either negative or positive, or lines that are horizontal or vertical. A class discussion of the results will be a rich one, because it is likely that all of these situations will be addressed.

You know that a line goes through the point (4,2) and that it slants up and to the right. Name at least one other thing that you are sure is NOT true about the line.

This question is accessible to any student. A student who is unable to visualize the line could simply use a piece of graph paper, plot (4,2), and draw a line that slants up and to the right. Asking what is *not* true is actually a bit more interesting than asking what is true.

Some possible responses are: the line does not go through (4,1), the line does not have a negative slope, or the line does not go through both **Quadrants** II and IV.

Variations. Instead of using a line, the question could specify an exponential function that goes through (5,–5) and (8,73) or a quadratic that goes through (0,0) and (1,8). Or the conditions could specify a different type of feature of the graph than a point it goes through, for example, indicating that it is a **periodic function** that has an amplitude of 10.

Open Questions for Grades 9–12

> Two lines are **perpendicular** to each other and intersect at the point (10,5). What might be the slopes of these lines? Explain your thinking.

This question is open to many possible answers. Students can make a graph by hand or using technology to represent the possible perpendicular lines and the intersection point. If they do not know the relationship of slopes for perpendicular lines, they will discover this when determining the slopes of their particular lines.

Variations. An alternate formulation of the question could specify that the two lines are **parallel** and have y-intercepts at particular points, for example, (0,5) and (0,10).

> Rebecca said that you must know 3 points on a parabola to graph it. Aaron said that you only need to know 2 points and Tara said that you must know 4 points. Who is right? How do you know?

Students often know that two points determine a line but are sometimes less familiar with the notion that three points determine a parabola. However, if additional information is known about the points, the number required could change.

Some students will select pairs, triples, or sets of four points and try to graph parabolas that go through them. Others will also preselect sets of points but will work more algebraically to determine a parabola that goes through them. Still other students might use algebraic reasoning to solve a system of equations in three unknowns (the values of a, b, and c in $y = ax^2 + bx + c$).

Some students might realize that if it is known that a particular point is the vertex, only two points are required to establish the equation of the parabola. For example, if the vertex is (0,0) and it goes through (1,4), then the only parabola possible is $y = (x + 1)^2$. Other students might decide that if they are given the right four points there could be a parabola that goes through all of them, but there might be no single parabola that goes through an arbitrary four points.

> Write an algebraic expression involving one variable and at least three terms that can be simplified by using **exponent laws**.

There are many possible responses to this question partially because there are several laws of exponents and partially because there is variety possible using any one of the laws. For example, a student might choose $(4^2)^3 + x - 3$ or might choose $(x + 2) + (x + 2)^3 + [(x + 2)^2]^2$.

A quadratic equation of the form $x^2 + 10x + c$ can be factored in the form $(x + a)(x + b)$, where a and b are whole numbers.

Another quadratic equation of the form $x^2 + 12x + d$ can be factored in the form $(x + e)(x + f)$, where e and f are whole numbers.

Which do you think is greater: c or d? Why?

Although we often ask students to factor individual quadratics, we rarely ask them to compare the results of two factorings simultaneously. This question is interesting in that some students will think c is greater (e.g., if a and b are 5 and 5, and e and f are 10 and 2, then c is greater) and some will think d is greater (e.g., if a and b are 10 and 1, and e and f are 6 and 6, then d is greater). By asking many students their responses, students might even explore how often one value is greater than another.

A teacher might expect the answer "You can't be sure." In this case, given the wording of the question, an answer of c or d is equally correct.

For students who do not know how to begin, questions such as these could be helpful:

- *Could a be 12? Why or why not?*
- *What do you know about how a and b relate?*
- *What happens to d if e and f are close?*

A **periodic function** has a period of 6 and an amplitude of 3. Sketch a graph of this function.

This question allows students to sketch graphs of any periodic function with period 6 and amplitude 3. Students have the choice of how they wish to work in the periodicity of their graph, not restricting them to a just a **sine** or **cosine** type graph.

✳ BIG IDEA. **The principles and processes that underlie operations with numbers and solving equations involving numbers apply equally to algebraic situations.**

When you add a polynomial that can be represented with 3 algebra tiles to one that can be represented with 5 algebra tiles, how many tiles does the sum require?

Students can experiment to see that the number of tiles in the sum could actually range from 2 tiles to 8 tiles, but only even numbers of tiles.

If students struggle to come up with a response, the teacher could ask scaffolding questions such as:

- *Could it take 8 tiles? What might that possibility be?*
- *Could it take 10 tiles? Why not?*
- *Could it take only 2 tiles? How would that happen?*

Variations. The numbers of tiles in the addends could change. Subtraction or multiplication could be used instead of addition.

> Two of the roots of a **quartic equation** $ax^4 + bx^2 + c = 0$ are **complex conjugates** of each other. What can this quartic equation be? What are its roots?

Quartic equations are fourth-degree polynomial equations. In this case, students might observe that the equation could be rewritten as $ay^2 + by + c = 0$, where $y = x^2$, and solve it using what they know about quadratic equations.

Knowing that two of the roots are $a + bi$ and $a - bi$ impacts how students proceed. Some students might choose two of the roots to be i and $-i$ to simplify their work; others will use more complex situations. For example, a student might decide that the quadratic equation in y could be $y^2 + 1 = 0$, since i and $-i$, which are complex conjugates, are solutions. Since $y = x^2$, that means that $x^2 = i$ or $-i$. Solving for x yields

$$\frac{1+i}{\sqrt{2}}, \ \frac{-1+i}{\sqrt{2}}, \ \frac{1-i}{\sqrt{2}}, \ \frac{-1-i}{\sqrt{2}}$$

Other students might choose more complicated equations to work with.

TEACHING TIP. A question that asks students to come up with an equation for certain roots is more accessible to students because they can work "backward" rather than having to address a question for which they do not feel they have an obvious starting point.

✳ **BIG IDEA.** The transformations that are fundamental to determining the geometric relationships between shapes apply equally to algebraic situations.

> The function $f(x) = \sin x$ is transformed by **translating** to the right and then vertically **stretching**. Draw the new function on the graph. What is the equation of the new function?

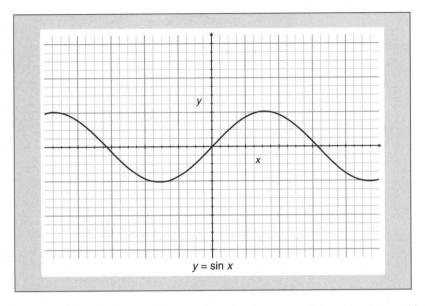

$y = \sin x$

Students have choice in this open question in terms of the size of the shift to the right and the amount of vertical stretch. Some students may draw the new graph on the same graph as the sine function if this helps them see the transformations more easily. Other students may decide what the new equation would be and draw the graph right away.

For students having difficulty getting started, questions that could be asked include:

• *How does the shift to the right affect the points on the y = sin x graph?*
• *Does the graph move up or down? How do you know?*
• *How does the vertical stretch affect the amplitude? The period?*
• *How does the original graph of y = sin x help you visualize the new graph?*

> You started with a parabola, then applied a transformation. The new parabola is **congruent** to the original one, but it is moved to the right and up.
>
> If you wrote the two equations, what would you expect to be different about them? What would you expect to be the same?

Some students will graph the parabola and actually determine the equations. Others will be able to predict which of the parameters of the parabola are likely to change and which are not. They could consider axis of symmetry, roots, values of the coefficients of x^2 and x in standard form, or consider the **factored form** or vertex form.

Variations. Rather than being congruent, the parabola might be moved straight up but be wider.

> A parabola has an axis of symmetry of $x = 3$. What might its equation be? How could you transform this parabola so that the axis of symmetry would be at $x = -3$?

Information about the axis of symmetry of a parabola tells a lot about it, but not everything. Students can move the parabola by stretching, **reflecting**, and translating as many times as they want. The only condition for the transformations is that the final position must have an axis of symmetry of $x = -3$.

If students have difficulty getting started, the teacher could provide scaffolding with questions such as these:

- *What does the line representing the axis of symmetry look like on the graph?*
- *Could the y-intercept for the parabola be 0? What could it be?*
- *What might the equation of the first parabola be? How did you figure it out?*
- *Why might it be useful to write the equation of the parabola in vertex form to get the second parabola?*

Variations. Different features of the parabola could be given, and a different transformation could be requested. For example: A parabola has a minimum point and no *x*-intercepts. How could you transform this parabola so that there would be a maximum point and a *y*-intercept of 4?

PARALLEL TASKS FOR GRADES 6–8

> **PARALLEL TASKS** are sets of two or more related tasks that explore the same big idea but are designed to suit the needs of students at different developmental levels. The tasks are similar enough in context that all students can participate fully in a single follow-up discussion.

✳ **BIG IDEA.** **Algebraic reasoning is a process of describing and analyzing generalized mathematical relationships and change using words and symbols.**

> Can 3,087 be in the pattern described by the given pattern rule? How do you know?
>
> <u>*Option 1*</u>: The pattern rule is: *Start at 9. Keep adding 3.*
>
> <u>*Option 2*</u>: The pattern rule is: *The term value is 4 times the term number + 3.*

Both of these options have students consider how the general term of a pattern describes the elements of that pattern. In <u>*Option 1*</u>, the rule is recursive, where each term is defined in terms of the preceding one.

In **_Option 2_**, an explicit formula relating term number and term value is used. Many students who are less comfortable creating pattern rules based on term numbers are more comfortable applying those rules.

Whichever option is chosen, students are considering numerical properties that influence whether or not a particular number can be in the pattern. Questions applicable to both tasks include:

- *Would you need to perform a lot of calculations to decide whether 3,087 belongs in the pattern? Explain.*
- *What properties of 3,087 might be relevant in deciding whether it belongs in the pattern?*
- *If you knew that 3,084 was in the pattern, would it make the question easier to answer? Explain.*
- *Do you think 3,087 belongs? Why or why not?*

Variations. Students could be given different pattern rules or be asked about whether a different term value belongs in the pattern.

Write each of the three phrases as an algebraic expression.

Option 1:
- Triple a number
- Subtract a number from 10
- Add a number to itself

Option 2:
- Add 1 to a number and then double it
- Subtract a number from 10 and then divide by 4
- Triple a number, add 2, and then add 1 more

In both of these situations, the student is working on the relationship between verbal expressions and algebraic ones. In **_Option 1_**, students work with phrases where only one operation is used; in **_Option 2_**, more than one operation is used.

Regardless of the option selected, follow-up questions such as these could be asked:

- *What operation sign(s) appear in your expression? Why those?*
- *Are there other ways to write your expression algebraically? How do you know?*
- *Did it matter which letter you used for the variable? Why or why not?*
- *Which expression was easiest for you to translate? Why was that?*
- *Could the words have been slightly different but your expression the same? Explain.*

Variations. Students could be given different phrases to write as algebraic expressions, or could be given different algebraic expressions and asked to write them in words, or a combination of the two.

✳ **BIG IDEA.** Comparing mathematical patterns or relationships either algebraically or graphically helps us see that there are classes of relationships with common characteristics and helps us describe each member of the class.

> Create two patterns that are very much like the given pattern. Explain why the three patterns are alike.
>
> *Option 1:* 3, 4, 6, 9, 13, 18
>
> *Option 2:* 3, 7, 11, 15, 19

The difference between the two options is in the complexity of the given pattern. In *Option 2*, the pattern is a simple linear growing pattern, increasing by the same amount each time, while in *Option 1*, the pattern is somewhat more complicated, with the increase changing each time.

Relevant questions for both tasks include:

- *What makes the given pattern a pattern?*
- *Do you think the patterns need to start with the same number to be alike?*
- *How might that influence your decision about what pattern might be like it?*
- *What sort of pattern do you think would NOT be like the given one?*

Variations. Instead of giving patterns in numerical form, patterns of figures might be used instead.

TEACHING TIP. By asking students to create an item like an existing one, they are likely to consider many attributes of the provided item. The openness is in allowing them to choose what attribute to consider.

> You graph the points described in each **table of values**. How are the graphs alike and different?
>
> *Option 1:*
>
> | x | 1 | 2 | 3 | 4 |
> |---|---|---|---|---|
> | y | 4 | 7 | 10 | 13 |
>
> | x | 1 | 2 | 3 | 4 |
> |---|---|---|---|---|
> | y | 9 | 17 | 25 | 33 |
>
> *Option 2:*
>
> | x | 1 | 2 | 3 | 4 |
> |---|---|---|---|---|
> | y | 5 | 8 | 11 | 14 |
>
> | x | 1 | 2 | 3 | 4 |
> |---|---|---|---|---|
> | y | 9 | 12 | 15 | 18 |

The difference between the two options is in the complexity of the given patterns. Students are likely to notice that in *Option 2*, the values in each table go up by the same amount and in **_Option 1_**, they do not. When students plot their graphs, they might notice that the lines for **_Option 1_** both go through (0,1) and that the lines for **_Option 2_** are parallel.

No matter which option students choose, follow-up questions such as these would be appropriate:

- *How are your graphs similar? How are they different?*
- *What other table of values might result in a graph similar to the two you have?*

✳ BIG IDEA. **The same algebraic expression or pattern can be related to different situations, and different algebraic expressions can describe the same real-world situation, sometimes with limitations based on context.**

> **_Option 1:_** A rectangle has a length that is 3 units greater than its width. What algebraic expression describes its perimeter?
>
> **_Option 2:_** The expression $3x + 3$ describes the perimeter of a shape. What might the shape be? How do you know?

Option 1 requires students to consider how to represent the various dimensions of a rectangle algebraically. The student might represent the length as $x + 3$, the width as x, and the perimeter as $4x + 6$. Alternatively, the student might represent the width as $x - 3$, the length as x, and the perimeter as $4x - 6$.

Option 2 requires the student to think about what shape might be possible and then determine various dimensions. It might be a triangle with side lengths x, x, and $x + 3$, or it might be a rectangle with length $\frac{3}{2}x$ and width $\frac{3}{2}$. There are other possibilities as well.

Questions applicable to both tasks include:

- *Could two of the sides of your shape have the same length? Why or why not?*
- *Could all the sides of your shape have the same length? Why or why not?*
- *Do you have to use a variable to describe each side length or only some of them? Why or why not?*
- *Describe all of the side lengths of your shape.*

✳ **BIG IDEA.** **Many equivalent representations can describe the same pattern or generalization. Each representation may give more insight into certain characteristics of the situation or generalization.**

> Draw a picture for the pattern that would help someone figure out what the 100th number in it would be.
>
> **Option 1:** 1, 3, 5, 7, 9, . . .
>
> **Option 2:** 4, 7, 10, 13, 16, . . .

In either situation, the student needs to consider how the picture helps show the growth in the pattern in some simple way. It might be, for **Option 1**, that the student draws something like the picture below, making it clear that the number in position n is twice $(n-1) + 1$.

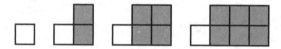

It might be, for **Option 2**, that the student draws something like the picture below, making it clear that the number in position n is $3n + 1$.

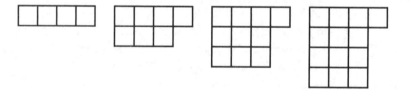

Regardless of the option selected, follow-up questions such as these could be asked:

- *What was the repetition in your pattern?*
- *How did your diagram show that repetition?*
- *How does your diagram make it easy to see what the 10th number in the pattern would be?*
- *How does it make it easy to see the general term?*

Variations. Students could be provided with alternative patterns to model.

✴ **BIG IDEA.** Limited information about a mathematical pattern or relationship can sometimes, but not always, allow us to predict other information about that pattern or relationship.

> You apply a consistent change rule to a lot of numbers. You want to know how the number 100 will be changed if ...
>
> _Option 1:_
>
> 10 becomes 25
> 20 becomes 45
> 40 becomes 85
>
> _Option 2:_
>
> 10 becomes 76
> 19 becomes 148

Each of these tasks has the students consider the notion that algebra is a way to describe a consistent change. In the first instance, more data are provided, the numbers used are simple to calculate with, and the rule involves addition rather than subtraction.

Relevant questions for both tasks include:

- _Do you think the resulting number will be more than 100? Why?_
- _Do you think it will be more than 200? Why?_
- _Do you think you can use the "evenness" or "oddness" of the first original number to predict the "evenness" or "oddness" of the result?_
- _Without calculating a value, what would you estimate the change result for 15 to be? Why?_
- _What was your prediction? How did you figure it out?_

Variations. More data can be provided for one of the options. The rule used might be much simpler (e.g., tripling the original number) or it might be much more complex (e.g., involving squaring the number and then performing further operations with the squared value).

✴ **BIG IDEA.** The principles and processes that underlie operations with numbers and solving equations involving numbers apply equally to algebraic situations.

> Show a simpler way to describe the expression. How do you know the simpler way is equivalent to the original?
>
> _Option 1:_ $4 - 3x + 17 + 9x$
>
> _Option 2:_ $8y + 9x - 7x^2 - 2y + 6 - 3x$

This task is designed to highlight the collection of like terms using familiar rules of integer operations.

No matter which option students choose, follow-up questions such as these would be appropriate:

- *Why did you say there were 6x?*
- *Why did only some of your terms go together but not all of them?*
- *How could you have predicted how many different terms you would end up with?*
- *What did you have to know about addition and subtraction to rewrite your expression?*

PARALLEL TASKS FOR GRADES 9–12

✳ **BIG IDEA.** **Algebraic reasoning is a process of describing and analyzing generalized mathematical relationships and change using words and symbols.**

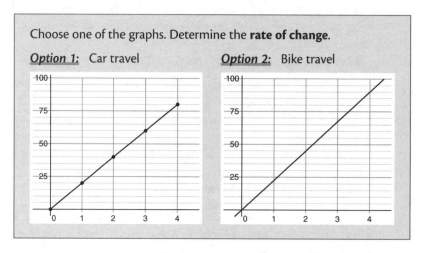

The numbers used to determine the rate of change are simpler in **Option 1** and the points are clearly defined on the line; both of those traits make it easier to calculate the **rise** and **run** of the graph and, therefore, the rate of change. The graph for **Option 2** has no points and allows the student to choose their own two points on the line to determine the rate of change.

Questions applicable to both tasks include:

- *What information do you need to calculate the rate of change from the graph?*
- *How could you have predicted whether the rate would be positive or negative before you did any calculations?*
- *Which points on the line did you choose to help you calculate the rate of change? Why did you choose those points?*
- *How did you use the points you chose? What steps did you have to take to determine the rate of change?*
- *What does the rate of change represent in this situation?*

Find the equation of the line that completes the shape:

Option 1:	Option 2:
A **parallelogram**	A right triangle
$y = 8$	$y = -2x + 8$
$y = -3x + 12$	$y = \frac{1}{3}x$
$y = 2$	

Both options require students to determine the equation of a line that would complete the shape, and in both cases there are many solutions. However, students need to bring geometry knowledge to bear. They need to know, for the parallelogram, that parallel lines have the same slope and, for the triangle, that the lines are perpendicular and thus have slopes that are negative reciprocals. Having two options allows students to start on the problem they are more comfortable with; work on either option can begin with graphing by hand or with technology.

Regardless of the option selected, follow-up questions such as these could be asked:

- *Did you use the slope information or the intercept information from the given equations to help you?*
- *Is there more than one solution?*
- *Could you have found the solutions without a picture?*

TEACHING TIP. Using questions that relate content strands is both practical and meaningful for students.

✳ **BIG IDEA.** Comparing mathematical patterns or relationships either algebraically or graphically helps us see that there are classes of relationships with common characteristics and helps us describe each member of the class.

Graph the rational function.

Option 1:	Option 2:
$f(x) = \frac{2x - 1}{x - 2}$	$f(x) = \frac{2x^2 - 2x + 3}{x - 2}$

Although both options involve graphing a rational function, the student can choose from two functions with different complexity levels. In **Option 2**, the degree of the polynomial in the numerator is greater than in the denominator; this may be more challenging for some students because it leads to an **asymptote** that is neither vertical nor horizontal, but is the line $y = 2x + 2$.

Relevant questions for both tasks include:

- *What is the **domain** of your function?*
- *Where is the vertical asymptote and horizontal or **slant asymptote** for your function?*
- *Did you find the −x and −y intercepts to help you graph your function?*

> Another function is a lot like the given function. What might it be? Why are the two alike?
>
> <u>Option 1:</u> $y = 3x^2 + 4$
>
> <u>Option 2:</u> $y = 3^x$

Both tasks require students to think about what makes relationships alike. They might choose to focus on either the algebraic representations or the graphical ones. In follow-up discussion they could be encouraged to consider the type of representation they did not try initially.

No matter which option students choose, follow-up questions such as these would be appropriate:

- *What happens to your function when x gets big? When x is quite small?*
- *What shape would the original function looked like if you graphed it?*
- *What kinds of things might you change but still have a similar function?*
- *What would you choose not to change?*

Variations. Instead of giving functions in algebraic form, the functions could be given in graphical form instead.

✳ **BIG IDEA.** The same algebraic expressions can be related to different situations, and different algebraic expressions can describe the same real-world situation, sometimes with limitations based on context.

> <u>Option 1:</u> Juan is planning a party. The restaurant charges $60 for the room and $10 per person for food. The community center charges $200 for the room and food for up to 16 guests, and then charges an additional $6 per person. To get the best price, which location should Juan use for his party?

> ___Option 2:___ Eric is planning a party. The restaurant charges $60 for the
> room and $10 per person for food. The community center
> charges $150 for the room and food for up to 10 guests,
> and then charges an additional $6 per person. If there are
> more than 20 guests, the community center charges $200
> instead of $150 for the hall rental. The food charges do
> not change. To get the best price, which location should
> Eric use for his party?

In both cases, one location offers a rate that relates directly to the number of guests and a second location offers a flat fee or an additional fee per person depending on the total number of guests. ___Option 2___ includes another change in charges after 20 guests. For either option, students can begin by creating either tables or graphs to help understand the different choices and the best value.

Questions applicable to both tasks include:

- *What do you need to consider when deciding which location to use for the party?*
- *Did the graphs show linear relationships between the cost and the number of people? How do you know a graph is or is not linear?*
- *Did the graphs intersect for the two locations?*
- *What would be the best choice for 10 people? 15? More than 20? Explain.*

> ___Option 1:___ Provide an example where the domain of a function used
> to model a real-world situation must be restricted. Explain
> why this would be necessary.
>
> ___Option 2:___ Provide an example where the **range** of a function used
> to model a real-world situation must be restricted. Explain
> why this would be necessary.

When using functions as models for real-world situations, the domain and range may be restricted to values that make sense for the situation. This set of parallel tasks gives students the option of considering restrictions on either the domain or range, as well as allowing them to choose the type of function. Students can (a) think of a context where a function is used to model the situation; (b) decide whether the function is linear, quadratic, or sinusoidal; and (c) decide which domain and range values need to be excluded (e.g., negative heights or negative times).

Regardless of the option selected, follow-up questions such as these could be asked:

- *Is your function linear, quadratic, or sinusoidal?*
- *Did a graph of your function help you with the question?*

- *What real-world context did your function model?*
- *Why does it make sense that the graph was restricted?*
- *What is the domain or range of your function?*
- *How did you decide what the restrictions were for your domain or range?*

A ball is thrown, and the height (in meters) as a function of time (in seconds) is $h = -5t^2 + 20t + 2$.

Option 1: For how many seconds was the height of the ball at least 17 meters?

Option 2: How long was the ball in the air?

These parallel tasks ask students to work with a projectile problem that is modeled with a quadratic equation. The questions may seem similar, but **Option 1** is more challenging than **Option 2**, where students only need to find the zeroes. Both options allow students to make connections between the quadratic equation and how the height of the ball changes with time. Maximum height, distance from the ground when the ball was thrown, and time until the ball hits the ground can all be drawn out in the class discussion.

Relevant questions for both tasks include:

- *Why does it makes sense that the function is a quadratic and not linear?*
- *If you graphed your function, what would it look like? How do you know?*
- *When is the ball at its highest point?*
- *When does the ball hit the ground?*
- *What was your solution, and what strategy did you use?*

Option 1: A T-shirt printing company has start-up costs of $750 and a cost of $8.50 for each printed T-shirt. Use a graph of the rational function to show the average cost of producing *n* T-shirts.

Option 2: A T-shirt company bought a case of T-shirts for $750. Five of these shirts were kept for displays and the rest sold for $1,280, with a profit of $13.44 per shirt. Use a graph of the rational function to show how many T-shirts were in the case.

Both options use a rational function to model the T-shirt sales, but for different purposes. **Option 1** sets up a rational function to look at the average cost of producing T-shirts and see what happens as the number of T-shirts produced becomes larger. This option may be more direct for the student. **Option 2** sets up a rational function to model the sale of T-shirts and find the zeros from the graph to determine the original number of T-shirts in the case. Both options require students to consider inadmissible solutions that are outside the domain determined by the context of the problem.

No matter which option students choose, follow-up questions such as these would be appropriate:

- *What is the rational function you used to solve your problem?*
- *What was the domain in the context of your problem? Explain.*
- *What was the equation of your vertical asymptote?*
- *Did your graph have an x-intercept?*
- *How did you solve your problem using the graph?*

✳ **BIG IDEA.** Many equivalent representations can describe the same pattern or generalization. Each representation may give more insight into certain characteristics of the situation or generalization.

> Use algebra tiles to model your polynomials.
>
> **Option 1:** Two polynomials that add to $6x^2 + 8x + 2$.
>
> **Option 2:** Two polynomials that multiply to $6x^2 + 8x + 2$.

Even though many teachers would not consider dealing with polynomial multiplication and addition in the same lesson, there may well be students not ready for the multiplication problem who could be successful with the addition. What the problems have in common is that students consider alternatives for representing algebraic operations. In particular, they might consider the zero principle, or they might consider why tiles need to be more particularly arranged for multiplication (e.g., in an array).

Questions applicable to both tasks include:

- *What algebra tiles show $6x^2 + 8x + 2$? Is there any other way to model that same polynomial?*
- *How did you arrange your algebra tiles? Is there any other way you could have arranged your tiles?*
- *Where did you start? Why did you start there?*

Variations. Students might be told that an area is doubled or quadrupled and asked what happens to the radius, the **circumference**, or the diameter.

TEACHING TIP. It is always useful to explore "what if" questions—where one assumption or constraint in a situation is altered to see its effect on other aspects of the situation.

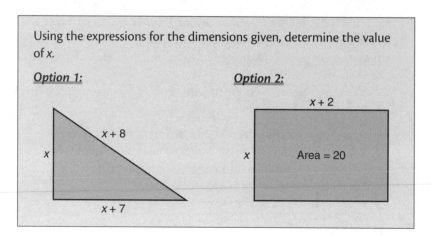

Both options require the student to set up a quadratic equation and to find the factors to determine the value of x that is required. With **Option 1**, the student needs to use the **Pythagorean theorem**. **Option 2** is more straightforward: the student calculates the given area by multiplying the expressions for length and width.

Regardless of the option selected, follow-up questions such as these could be asked:

- *What properties of your shape did you use to help you find the value of x?*
- *How did your equation help you find the value of x?*
- *Did you verify your value of x? How?*

Option 1:   Using color tiles, create a pattern representing an arithmetic sequence and one representing a **geometric sequence**. What is the sum of the first 25 terms of each of these sequences?

Option 2:   Using color tiles, create a pattern that simultaneously represents an arithmetic sequence and a geometric sequence when different attributes are considered. What is the sum of the first 25 terms of this sequence?

When students are building on their understanding of arithmetic and geometric sequences, the similarities and differences between these types of sequences can be enhanced with concrete representations. For both options, students find the sum of a series for 25 terms, although **Option 1** is less complex.

Relevant questions for both tasks include:

- *How did you use the color tiles to represent your arithmetic pattern? Your geometric pattern?*
- *What makes the arithmetic pattern different from the geometric one?*
- *What is the nth term for your arithmetic and geometric pattern?*
- *Which pattern grows faster? Why?*
- *How did you find the sum of the first 25 terms for each pattern?*

✳ **BIG IDEA.** Limited information about a mathematical pattern or relationship can sometimes, but not always, allow us to predict other information about that pattern or relationship.

> _**Option 1:**_ A line of slope $\frac{2}{3}$ goes through (−4,−1). Name two more points on the line.
>
> _**Option 2:**_ A line of slope $-\frac{2}{3}$ goes through (−4,−1). Name two more points on the line.

Some students find it difficult to work with negative slopes, so allowing them the chance to work with _**Option 1**_ will maximize the potential for success. In each case, students deal with the fact that if you know one point on a line and the slope of a line, you know a lot more, too.

No matter which option students choose, follow-up questions such as these would be appropriate:

- _Do you know which way your line slants? How do you know?_
- _Could (−4,1) be on your line? How do you know?_
- _Could (−3,0) be on your line? How do you know?_
- _Why was it a good idea to choose one point where the x-coordinate was 3 away from −4?_
- _What points did you use?_

> Consider only situations where $x > 0$.
>
> _**Option 1:**_ Graph the following lines on the same grid and identify the coordinates of the point of intersection:
>
> $$y = x \qquad y = x + 3$$
>
> _**Option 2:**_ Graph the following lines on the same grid and identify the coordinates of the point of intersection:
>
> $$y = 3x \qquad y = \tfrac{1}{2}x$$

Both options investigate the conditions where the lines never cross once x is greater than 0. With one option, students will work with parallel lines and with the other option students will work with two lines that start from the origin and increase linearly.

Questions applicable to both tasks include:

- _Where do your lines intersect on the graph?_
- _Will they ever intersect?_
- _What properties of the lines ensure that the lines do not intersect for positive values of x?_
- _What could you change to make the lines intersect?_

✳ **BIG IDEA.** The principles and processes that underlie operations with numbers and solving equations involving numbers apply equally to algebraic situations.

Option 1: Substitute various values for x into the expression

$$x^2 + 2x + 1$$

Factor the numbers you get. How could that help you factor $x^2 + 2x + 1$?

Option 2: Substitute various values for x into the expression

$$3x^2 + 7x + 2$$

Factor the numbers you get. How could that help you factor $3x^2 + 7x + 2$?

Both options help students build a connection between factoring algebraic expressions and factoring numbers. For example, a student doing **Option 1** would notice that each time, the values $(x + 1)$ and $(x + 1)$ were being multiplied.

x	$x^2 + 2x + 1$	Factored value
1	4	2·2
2	9	3·3
3	16	4·4
4	25	5·5

Regardless of the option selected, follow-up questions such as these could be asked:

- *What does it mean to factor numbers?*
- *What does it mean to factor polynomials?*
- *How did it help to look for a pattern in the factored values?*
- *Once you saw the pattern, why did you relate what you saw back to the value of* x *in the row?*

✳ **BIG IDEA.** The transformations that are fundamental to determining the geometric relationships between shapes apply equally to algebraic situations.

The equation of a sine function is $y = a \sin b(x - d) + c$.

Option 1: If the graph passes through the origin, what do you know about a, b, c, and d?

Option 2: If the graph has no x-intercepts, what do you know about a, b, c, and d?

Parallel Tasks for Grades 9–12

Students are familiar with questions in which they are given a trigonometric function and asked to graph the function. Although the difficulty levels of **_Options 1_** and **_2_** are not significantly different, the fact that students can choose which option to work on will help many students get started.

Relevant questions for both tasks include:

- *Did you use a graph of sin* x *to help you with the task?*
- *Were all variables* a, b, c, *and* d *affected?*
- *What transformations of the sin* x *graph are involved?*

TEACHING TIP. Providing choice, whatever the choices may be, is one important way to differentiate instruction.

> **_Option 1:_** Two trigonometric functions are graphed. It looks like one was just shifted to the right. What could the equations be?
>
> **_Option 2:_** Two trigonometric functions are graphed. One has a lot more "bumps" in the same space than the other, but it's no taller. What could the equations be?

Both options require the student to consider the shape of sinusoidal graphs and the equations that can describe their shape. In one case the student does a horizontal shift, whereas in the other case there is a horizontal **compression**, which some students find more complex.

No matter which option students choose, follow-up questions such as these would be appropriate:

- *What form might the equations of the graphs have? Why do you think that?*
- *How might the equations of your two graphs differ?*
- *How did looking at the transformation that was applied help you to figure out how the equations might differ?*
- *How else could you transform your graph? How would the equation change that time?*

SUMMING UP

The seven big ideas that underpin work in Algebra were explored in this chapter through 82 examples of open questions and parallel tasks, as well as variations of them. The instructional examples provided were designed to support differentiated instruction for students at different developmental levels, targeting two separate grade bands: grades 6–8 and grades 9–12.

MY OWN QUESTIONS AND TASKS
_____

Lesson Goal: Grade Level: _____

Standard(s) Addressed:

Underlying Big Idea(s):

_____

Open Question(s):

_____

Parallel Tasks:

Option 1:

Option 2:

_____

Principles to Keep in Mind:

- All open questions must allow for correct responses at a variety of levels.
- Parallel tasks need to be created with variations that allow struggling students to be successful and proficient students to be challenged.
- Questions and tasks should be constructed in such a way that will allow all students to participate together in follow-up discussions.

The examples presented in this chapter only scratch the surface of possible open questions and parallel tasks that can be used to differentiate instruction in Algebra, which is probably the most significant strand for grades 6 through 12. For example, there are many situations in which the teacher could provide limited information about a relationship (e.g., that its graph goes through (2,5), (6,37), and (9,82)) and ask students what else they might be able to conclude. Surely many new ideas have already come to mind. A form such as the one shown here can serve as a convenient template for creating your own open questions and parallel tasks. The Appendix includes a full-size blank form and tips for using it to design customized teaching materials.

Number and Operations

DIFFERENTIATED LEARNING activities in number and operations are derived from applying the NCTM process standards of problem solving, reasoning and proof, communicating, connecting, and representing to content goals of the NCTM Number and Operations Standard, including

- understanding numbers, ways of representing numbers, relationships among numbers, and number systems
- understanding meanings of operations and how they relate to one another
- computing fluently and making reasonable estimates (NCTM, 2000)

TOPICS

Before beginning the task of differentiating student learning in number and operations, it is useful for teachers to have a good sense of how the topics in the strand develop over the grade bands. The NCTM *Curriculum Focal Points* (NCTM, 2006), which suggest what mathematical content should be the focus at each grade level through grade 8, were used as the basis for recommendations made in this resource for grades 6–8; the NCTM *Principles and Standards for School Mathematics* (NCTM, 2000) helped form the basis for the material for the higher grades. For a teacher at a particular grade level, it can be helpful to be aware of where students' learning is situated in relation to what learning has preceded the present grade band and what will follow.

Grades 6–8

Within this grade band, students extend their understanding of fractions and decimals to situations involving multiplication and division of these values. They increasingly work with ratio and **proportion**, particularly, but not exclusively, in percent situations. They also begin to work with more abstract values, including **negative integers**, **exponents**, and **scientific notation**.

Grade 9–12

Within this grade band, students become more comfortable with the relationship of sets within the number system including **rational**, **irrational**, and **complex numbers**, working with **powers** and exponents, counting using **permutations** and **combinations**, and working with **matrices**.

THE BIG IDEAS FOR NUMBER AND OPERATIONS

Coherent curricula in number and operations that meet NCTM content and process standards (NCTM, 2000) and support differentiated instruction can be structured around the following big ideas:

- Numbers tell how many or how much.
- Classifying numbers provides information about the characteristics of those numbers.
- There are many equivalent representations for a number or numerical relationship. Each representation may emphasize something different about that number or relationship.
- Numbers are compared in many ways. Sometimes they are compared to each other. Other times, they are compared to **benchmark** numbers.
- The operations of addition, subtraction, multiplication, and division hold the same fundamental meanings no matter the domain to which they are applied. Each operation has an **inverse operation**.
- There are many **algorithms** for performing a given operation.

The tasks set out and the questions asked while teaching number and operations should be developed to evoke these ideas. The following sections present numerous examples of application of open questions and parallel tasks in development of differentiated instruction in these big ideas across two grade bands.

OPEN QUESTIONS FOR GRADES 6–8

OPEN QUESTIONS are broad-based questions that invite meaningful responses from students at many developmental levels.

✸ BIG IDEA. **Numbers tell how many or how much.**

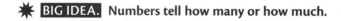

Jeff says that −1,000 might be a lot, but Ian says it's not. What do you think? Why?

This very open-ended question uses language that is deliberately vague, but at the same time it piques curiosity because it suggests that there could be a disagreement about this mathematical statement. It also promotes student math talk, which is something to be valued in the math classroom.

If students do not know how to begin, the teacher could provide scaffolding with questions such as these:

- *How could you model −1,000?*
- *Would your model give a sense of its size or not?*
- *How would your model give a sense of the size of −1,000, or why would it not do that?*
- *How are 1,000 and −1,000 the same? Different?*
- *Do you think of 1,000 and −1,000 as the same size or different sizes?*

Variations. Rather than using positive and negative values, it is possible to use fractions and ask, for example, whether $\frac{2}{3}$ is a little or not.

TEACHING TIP. Using carefully vague language and allowing for mathematical differences of opinion that appear to be legitimate are both techniques for engaging student interest and willingness to participate. The care in this case was in choosing a situation that really could be looked at in different ways and in using language that made different viewpoints seem reasonable.

> What would be a good estimate for the number of leaves on a tree? Explain your thinking.

This question is an example of what is called a Fermi problem These problems require realistic estimation and consideration of **order of magnitude**. Typically, students must make assumptions in solving the problem, and stating those assumptions is an important part of the process. For example, in this situation, if a student uses the number of leaves on a single branch and the number of branches on a tree to help estimate the required value, the student needs to determine whether the "model" tree and "model" branch are typical.

The question is open in that there are many strategies a student could pursue to answer it. It is also accessible because the context on which it is based is one that is very familiar to students.

Variations. Rather than estimating the number of leaves on a tree, a student might estimate the number of pennies that could be fit on all the floors in the school, the number of people that would fill the school building if they stood close together, or the area that would be necessary for 6 or 7 billion people (the world population) to stand crowded together.

✴ **BIG IDEA.** **Classifying numbers provides information about the characteristics of those numbers.**

> You write a fraction as a decimal and it looks like 0.00☐☐.
>
> Tell three things that you are sure are true about the number.

Many students struggle with decimals involving thousandths or ten thousandths. This particular question is more accessible to them because it allows them, for example, to say that the number has four decimal places or five digits. It also allows them to say that the number is pretty small. However, it is also important that some students point out that the number is less than $\frac{1}{100}$ or that it could be written in the form $\frac{x}{10,000}$.

By asking students to tell something about either the fraction or the decimal, the question is made more open-ended. A teacher might encourage students to discuss both of those forms.

TEACHING TIP. As the teacher circulates while students are working, individual students who are ready can be prompted to offer more sophisticated responses that will benefit the class.

> You write a fraction as a percent and the percent is of the form
> ☐☐☐.☐%.
>
> What do you know about the fraction?

Because the exact value of the percent is not indicated, students can simply choose a simple value and write the corresponding fraction. For example, a student could choose something as simple as 200.0%, but would still need to write a fraction of the form $\frac{2}{1}$ or $\frac{20}{10}$ or, perhaps, $\frac{2,000}{1,000}$. Some students might write $\frac{200.0}{100}$, and there could ensue an interesting discussion about whether it is possible to have a decimal numerator within a fraction.

Other students might think more generally. For example, they could simply state that the fraction must be **improper** (since the percent is at least 100), that the fraction could be written with a denominator of 1,000 (since there is a decimal place in the percent), or that the fraction is greater than $\frac{1}{2}$.

✳ **BIG IDEA.** There are many equivalent representations for a number or numerical relationship. Each representation may emphasize something different about that number or relationship.

> Choose a very big number. Describe it in as many ways as you can.

This open question allows the teacher insight into what students think makes a number big. It also affords the opportunity for students to describe a number in many ways, whether it is using place value concepts, comparing it to other numbers, relating it to contexts where the number is reasonable, or explaining how a calculation might result in the number.

Because the students are allowed to choose the number, even students uncomfortable with certain types of numbers can still actively participate in the discussion. For example, a student might argue that 100 is big, if you're thinking of ages of people.

For students who struggle to come up with a response, questions such as these could be helpful:

- *Is 1,000 big? Could it be?*
- *Could the big number be a fraction or a decimal?*
- *How could you describe your number with a picture?*
- *How could you describe the number by relating it to other numbers?*

Variations. Rather than asking for a big number, the teacher could ask for a very small number instead.

> Fill in values for the blanks to make this statement true:
>
> _____ is $\frac{3}{5}$ of _____

By not simply asking students to determine $\frac{3}{5}$ of a particular number, this question encourages students to realize that there are many pairs of numbers in which one number is $\frac{3}{5}$ of another. In fact, there is an infinite number of responses.

Simple answers might be that $\frac{3}{5}$ is $\frac{3}{5}$ of 1 or that 3 is $\frac{3}{5}$ of 5. More complex responses might be that $\frac{3}{20}$ is $\frac{3}{5}$ of $\frac{1}{4}$ or that $\frac{8}{11}$ is $\frac{3}{5}$ of $\frac{40}{33}$. Some students might use verbal expressions instead, for example, "Watching sports is $\frac{3}{5}$ of my TV watching time."

By sharing responses, it is possible to bring to students' attention to the facts that in each strictly numerical description, if you multiply the second number by $\frac{3}{5}$, you get the first one, or if you divide the first one by $\frac{3}{5}$, you get the second one.

Variations. Rather than a fraction, a decimal or a percent could be used. For example, the problem could read "_____ is 25% of _____." Another variation could involve switching the positions of the blanks. For example, the expression could take the form "8 is _____ (fraction) of _____."

> Create a sentence that uses each of the following words and numbers.
> Other words and numbers can also be used.
>
> *0.6, little, decimal, 0.01*

This open question could produce a variety of responses. For example, students might write:

- *0.6 is a decimal that is fairly little, but 0.01 is even littler.*
- *If you multiply the decimal 0.6 by 0.01, you get a very little value.*
- *0.6 + 0.01 is a decimal that is only a little more than 0.6.*

The numbers 0.6 and 0.01 were chosen to encourage students to work with decimals rather than only whole numbers.

Variations. Values and words could be changed to prompt students to focus on other concepts or to work with more or less complex calculations. For example, to promote thinking about **proportional reasoning**, terms such as the following could be used: *30, 90, portion, fraction*. Or, the exercise might draw attention to powers or exponents with terms such as *25, power, less, 3*.

TEACHING TIP. The "build a sentence to include given words or phrases" strategy is useful in a wide variety of situations. Through judicious choice of the words and values that are to serve as the building blocks for their sentences, the teacher can lead students to consideration of a wide range of concepts and calculations of any desired degree of difficulty.

> Choose a **repeating decimal** that you could write as a fraction. Tell what the fraction is.

By allowing the student to choose the repeating decimal, the task becomes more accessible to struggling students. For example, the student might choose $0.1111\ldots$ if he or she happens to know that this is a decimal representation for $\frac{1}{9}$. Or the student might begin with a fraction and then create the repeating decimal.

The classroom discussion will make it clear that any repeating decimal can be written as a fraction. There might also be an interesting debate about whether, for example, $0.400000\ldots$ should be called a repeating decimal or not.

If students have difficulty getting started, the teacher could ask scaffolding questions such as:

- *What makes a decimal repeating?*
- *Suppose the instruction didn't say "repeating." Could a nonrepeating decimal be written as a fraction?*
- *Suppose you had figured out the fraction equivalent to the decimal 0.1111 . . . . How could you use that answer to help you get another one?*

✳ **BIG IDEA.** Numbers are compared in many ways. Sometimes they are compared to each other. Other times, they are compared to benchmark numbers.

> Choose a single value to replace both blanks: $\frac{4}{\Box}$ and $\frac{5}{\Box}$
>
> Name a fraction that is between the two fractions you created.

This question is open in that it allows students to choose the value for the denominator. Students who are more comfortable with decimals than fractions are likely to choose 10, creating $\frac{4}{10}$ and $\frac{5}{10}$, and name a decimal value such as 0.45 as occurring between them. Other students will choose the value 1 and use $4\frac{1}{2}$ or $\frac{9}{2}$ to be between 4 and 5.

Still other students will realize that *any* value can be used for the denominator. Such students might choose a value of, for example, 6 to create the first two fractions, and then use the **compound fraction** with $4\frac{1}{2}$ in the numerator and 6 in the denominator as the in-between fraction:

$$\text{A fraction between } \frac{4}{6} \text{ and } \frac{5}{6} \text{ is } \frac{4\frac{1}{2}}{6}.$$

These students might then go one step further and simplify the fraction by multiplying numerator and denominator by 2 (producing $\frac{9}{12}$ in this case).

The question allows for discussion of the notion that there are always fractions between any two fractions, while providing opportunities to investigate equivalence of fractions and other fraction concepts.

TEACHING TIP. Allowing students to choose a value in a problem often makes it more accessible to a wider range of students.

> What strategy would you use to compare $\frac{3}{17}$ and $\frac{6}{33}$?
>
> Would you use the same strategy to compare $\frac{1}{10}$ and $\frac{8}{9}$? Explain.

By allowing students to choose their own strategies, the question is opened up. Often a teacher will suggest a strategy, for example, using a **common denominator**, but this type of open question frees the student to self-determine a path. This question is designed, in fact, to make the traditional common denominator strategy acceptable but not the most efficient.

Some students might think of using a **common numerator** (e.g., writing $\frac{3}{17}$ as $\frac{6}{34}$ and realizing that $\frac{6}{34}$ must be less than $\frac{6}{33}$). By asking whether another comparison would be done in the same way, the door is opened to discussing even more strategies, for example, using benchmarks.

Open Questions for Grades 6–8

> Write two numbers in scientific notation that you think are really easy to compare when written that way (to tell which one is greater than the other).
>
> Then write two numbers in scientific notation that you think are harder to compare when written that way.
>
> Tell why you chose the pairs of numbers that you did.

Because students are allowed to choose the numbers to use, this question is very open. No matter what numbers students choose, they are likely to be thinking about how to compare numbers written in scientific notation and are, as a result, likely to think about why scientific notation is useful.

TEACHING TIP. Asking students their opinion on a matter is a valuable strategy to remove the risk for a student in responding to a question.

✺ BIG IDEA. The operations of addition, subtraction, multiplication, and division hold the same fundamental meanings no matter the domain to which they are applied. Each operation has an inverse operation.

> You add two numbers and the answer is negative. What two numbers might you have added?

The manner in which this question is stated allows a student to avoid the zero principle if that student is still not comfortable with it, but it also allows students comfortable with the zero principle to use that knowledge. For example, one student might respond with $(-3) + 0$, another with $(-1) + (-1)$, and a third with $(-5) + 3$. In the course of the discussion, it is likely that issues related to the zero principle will come out, for example, that $(-5) + 3$ is really $(-2) + (-3) + 3$, or $(-2) + 0$.

For students who do not know how to begin, questions that could be asked include:

- *What makes an integer negative?*
- *Could the two numbers you add both be positive? Explain.*
- *How could you use your knowledge that $3 + 4 = 7$ to help you answer this question?*

> Would you calculate $(-20) - (-3)$ the in same way you would calculate $20 - (-4)$? Explain your thinking.

In this situation, the student has a chance to show that it might be useful to use different subtraction meanings in different situations. For example, for the first question, it might be useful to think of taking away 3 negative counters from 20 negative ones, using the calculation $20 - 3$. In the other instance, the student might more likely think about adding the opposite, because going from -4 to 20 on a number line means adding 4 and 20.

Variations. The question can be varied to invoke different meanings of division. For example, a student is likely to calculate $(-21) \div (-7)$ by counting the number of groups of (-7) in (-21), but is likely to calculate $(-21) \div 3$ by sharing the (-21) into 3 equal groups.

> How is subtracting fractions like subtracting whole numbers? How is it different?

Students are free to focus on any aspects of subtraction they wish in answering this question. For example, one student might talk about the meanings of the operations, for example, both could mean take-away. Another student might focus on the kinds of answers you get, for example, a fraction answer when subtracting fractions and a whole number answer when subtracting whole numbers, if the lesser value is subtracted from the greater one. Yet another student might refer to the different sorts of models that are used in the two situations or the different levels of complication of the procedures.

Variations. The question can be varied by asking about other operations (e.g., addition, multiplication, or division) or by changing one or both sets of numbers to be compared (e.g., decimals or integers).

TEACHING TIP. Asking students how two items are alike, or different, is a useful strategy for creating open questions.

> How is thinking about what $\frac{4}{5} \div \frac{2}{5}$ means like figuring out what $\frac{4}{5} \div 2$ means? How is it different?

In this situation, students are not being asked what the answers to these two divisions are, but, instead, they are prompted to consider what the operations actually mean. For example, it is likely that they will think of the first situation as asking how many groups of $\frac{2}{5}$ are in $\frac{4}{5}$, but they are more likely to think of the second situation in terms of sharing $\frac{4}{5}$ into two equal groups.

The question is open because students are free to interpret the question in terms of what meanings of division might be called upon, but they also might simply say that the questions are alike because they are both divisions that start with $\frac{4}{5}$, and that they are different because $\frac{4}{5}$ is divided by different things. Although the latter approach to the question does not speak to how actually performing the calculations might be alike and different, it is a good starting point.

If students struggle to come up with a response, the teacher could provide scaffolding with questions such as these:

- *What does 6 ÷ 2 mean? Could that help you figure out what $\frac{4}{5} \div \frac{2}{5}$ means?*
- *What else could 6 ÷ 2 mean?*
- *Do you think of 18 ÷ 2 in the same way as you think about 32 ÷ 16? Why or why not?*

✳ **BIG IDEA.** **There are many algorithms for performing a given operation.**

> Imagine that you are planning to multiply the decimal 0.250 by a number. For what multiplication might you leave 0.250 as a decimal? For what number might you change it to a fraction? Explain your choices.

A question such as this encourages students to see that one might reasonably choose to use different procedures in different circumstances. At the same time, it is accessible because the students are free to make the case for whatever numbers they might choose.

> Which of these calculations do you see as most alike? Why?
>
> $$-3.4 + 5.7 = 1.7 \qquad \frac{3}{5} + \frac{2}{3} = 1\frac{4}{15} \qquad 3\frac{1}{2} + \left(-\frac{4}{3}\right) = 2\frac{1}{6}$$

When questions are posed in a manner that allows students to choose which items are most alike, they feel that their opinions are valued. This is important for building student confidence. In this case, some students are likely to think the last two calculations are most alike because both involve fractions. Others will think the first and last are most alike because they both involve positives and negatives.

As students justify their solutions, the teacher will learn a lot about their insights into number calculations.

Variations. Some students might be challenged to think of reasons that the last two computations are most alike (e.g., both involve working with thirds). Other sets of computations can be compared, or students might be asked to look for differences as well as (or instead of) similarities.

TEACHING TIP. Questions that ask students to choose which items in a list are most alike or most different allow student autonomy.

OPEN QUESTIONS FOR GRADES 9–12

✳ **BIG IDEA.** Numbers tell how many or how much.

> Choose an irrational number greater than 10. Square it. Estimate the value of the square and explain your estimate.

By providing students a choice of the irrational number to be used, the question is open. A student might choose the number $\sqrt{101}$, showing that he or she understands that 101 is not a perfect square; another student might choose, for example, 4π. Whatever choice is made, students need to be able to explain how they know the value is greater than 10 and then estimate the value of its square.

> Amy says that there is the same total number of positive and negative integers as the number of just positive integers. Ian says that makes no sense. What do you think? Why?

Most students will assume that there are more positive and negative integers than just positive integers. They will likely point out that the positive and negative integers together include all the positives, so the number of positive integers must be fewer. Other students might agree with Amy's suggestion, saying that both numbers are **infinity**.

It is the response to the question of how to be sure that can lead to a discussion about the fact that a one-to-one match can be created between the two sets. If there is a one-to-one mapping (as is shown below), the sets have to be equal in size. For example:

1 matches 1	2 matches 3	3 matches 5
−1 matches 2	−2 matches 4	−3 matches 6 ...

In other words, $+n$ gets matched with $2n - 1$, and $-n$ gets matched with $2n$.

Variations. Other infinite sets can also be compared, with students showing why the sets can be matched even when, on first blush, it looks like they cannot be. Examples are positive integers and positive integers greater than 100; even numbers and all whole numbers; or positive integers and fractions (this one being a much more difficult problem).

✹ BIG IDEA. **Classifying numbers provides information about the characteristics of those numbers.**

> The digits 23 repeat over and over to the right of the decimal point in a positive number. There are non-zero digits to the left of the decimal point. Tell some things you know about the number.

In being asked to tell some things about the number, a student is free to tell many things or just a few. Some students will realize the number is greater than 1, because there is a non-zero digit to the left of the decimal point. Some students will indicate that the fraction part of the number is greater than $\frac{2}{10}$ but less than $\frac{3}{10}$, or perhaps slightly less than $\frac{1}{4}$. Other students will realize that the fractional part of the number must be $\frac{23}{99}$.

TEACHING TIP. By asking students to tell some things about a number, rather than requiring a specific number of things, pressure on the struggling student is minimized.

> Devon says that he's thinking of a number that is a rational number when you multiply it by $\sqrt{2}$. What do you know about Devon's number?

To answer this, students must know the difference between rational and irrational numbers and must also realize that if you multiply a rational number by $\sqrt{2}$, the result is irrational and not rational. Beyond that, they have many options about what Devon's number might be or might not be.

For example, a student could say that Devon's number might be $\sqrt{2}$ or perhaps a rational multiple of $\sqrt{2}$, such as $3\sqrt{2}$. But it could not be $\sqrt{3}$ or a rational multiple of that number. Or a student might recognize that the number could be a rational multiple of any of $\sqrt{2}$, $\sqrt{8}$, $\sqrt{18}$, and so on, because each of these can be multiplied by $\sqrt{2}$ to get a rational answer (or they can be simplified to a rational multiple of $\sqrt{2}$).

Open Questions for Grades 9–12

✳ BIG IDEA. There are many equivalent representations for a number or numerical relationship. Each representation may emphasize something different about that number or relationship.

> Why might it be useful to write $5^4 \times 20^5$ in a different form to perform the calculation?
>
> Can you think of another pair of numbers that you might rename for a similar purpose?

The reason the exponent laws are used is to represent numbers in ways that simplify calculations. In the example provided, the numbers are simple enough for most students to work with. But they do make the point that combining numbers using exponent laws makes calculations simpler.

By asking students to think of another pair of numbers, the question is opened up even more. The responses also provide the teacher with an opportunity to assess student understanding.

> A number is written as a **root** of another number, but it's really easy to figure out the standard value without a calculator. What might the number be?

This question allows students to work at a level suited to their knowledge and comfort zone. Some students might use simple values, such as $\sqrt{100}$, whereas others might use values like $\sqrt{15^4}$ or $\sqrt{1{,}000{,}000{,}000{,}000}$ or $\sqrt[3]{27{,}000}$.

For students having difficulty getting started, questions such as these could be helpful:

- Are all **square roots** of powers of 10 easy to calculate without a calculator? Are some?
- What powers do you already know? How could that help you?

✳ BIG IDEA. Numbers are compared in many ways. Sometimes they are compared to each other. Other times, they are compared to benchmark numbers.

> An irrational number is about 8. What might it be?

This question is open because there are so many possible responses. Some include $\sqrt{65}$, $\sqrt{63}$, $\sqrt{68}$, $\sqrt{95} - 2$, and so forth. Most students will use the fact that

$8^2 = 64$ as a starting point, but that is not required. For example, a student who knows that π is about 3 might use the expression $2\pi + 2$, or a student who knows that $\sqrt[3]{27} = 3$ might use the expression $2\sqrt[3]{26} + 2$.

Variations. Instead of asking for an irrational number to compare to a rational one, a teacher could ask for a rational number that is estimated by a given irrational one. For example, one could ask for a rational number that is just a bit more than $\sqrt[3]{28}$.

✸ **BIG IDEA.** **The operations of addition, subtraction, multiplication, and division hold the same fundamental meanings no matter the domain to which they are applied. Each operation has an inverse operation.**

> How are adding and subtracting matrices like adding and subtracting real numbers? How are they different?

This question provides students the opportunity to show what connections they have made between operating with matrices and operating with numbers. Their responses will reveal understandings as well as misconceptions they might have.

If students do not know how to begin, the teacher could ask scaffolding questions such as:

- *What do you do when you add two numbers?*
- *What do you do when you add two matrices?*
- *In what situations do you subtract numbers?*
- *In what situations do you subtract matrices?*

Variations. Rather than focusing on adding and subtracting, the question could focus on either multiplying by a scalar or multiplying two matrices.

> Aline says it's possible to add $\sqrt{3} + 2\sqrt{3}$, or $\sqrt{3} + \sqrt{12}$, but not $\sqrt{3} + \sqrt{2}$.
>
> Do you agree or disagree? Explain.

In this situation, students are free to either agree or disagree. They might agree if they suggest there is no simplified form for $\sqrt{3} + \sqrt{2}$, but there is for the other sums. They might disagree by saying that the sum could be estimated in all cases. Ultimately, it is important for students to recognize that any two real numbers can be added by combining their values.

Variations. Alternate combinations of rational and irrational numbers could be proposed to be added.

✳ **BIG IDEA.** There are many algorithms for performing a given operation.

> Imagine that you are planning to subtract the rational number −0.750 from another number.
>
> For what subtraction might you leave −0.750 as a decimal?
>
> For what subtraction might you change it to a fraction?
>
> Explain your choices.

A question like this one encourages students to see that one might reasonably choose to use different procedures in different circumstances. At the same time, it is accessible because the students are free to make the case for whatever numbers they might choose. For example, they might decide to use a fraction form if subtracting from $\frac{1}{4}$, but a decimal form if subtracting from 3.21.

> Which of these calculations do you see as most alike? Why?
>
> $3\pi - 4\pi = -\pi$ $\qquad$ $8\sqrt{2} + 3\sqrt{2} = 11\sqrt{2}$ $\qquad$ $3\sqrt{8} + 4\sqrt{2} = 10\sqrt{2}$

It will be interesting to see which pair of calculations students see as most alike. They might choose the last two equations because both involve combining square roots or both involve addition. Or, because the square roots, on the surface, look different, they might instead choose the first and second equations, where whole number multiples of the same irrational are combined.

Variations. Some students might be challenged to think of reasons that the first and third computations are alike (e.g., both involve consecutive whole number multiples of an irrational). Other computations can be compared, and students can be asked to look for similarities and differences.

TEACHING TIP. Asking students to choose which items in a list are most alike or most different encourages them to study many attributes of those items.

PARALLEL TASKS FOR GRADES 6–8

> **PARALLEL TASKS** are sets of two or more related tasks that explore the same big idea but are designed to suit the needs of students at different developmental levels. The tasks are similar enough in context that all students can participate fully in a single follow-up discussion.

✳ **BIG IDEA.** **Numbers tell how many or how much.**

> *Option 1:* A positive number is a lot less than 0.04.
> What might it be?
>
> *Option 2:* A positive number is less than one millionth.
> What might it be?

Although both options involve students in considering the size of a fairly small number, one describes the number in words and the other in symbols. In the first situation, a student must decide what "a lot less" means. Because the number must be positive, the number must involve the use of fractions or decimals. In the second situation, the student could respond either symbolically or verbally, for example, 0.0000001 or, perhaps, 1 billionth.

Questions applicable to both tasks include:

- *Would you write your number as a fraction or as a decimal?*
- *Is 0.01 a possible value? Why or why not?*
- *Is 0.001 a possible value? Why or why not?*
- *What number did you choose?*
- *How could you write your number? How could you read it?*

Variations. Rather than small numbers, large numbers could be used instead. One number could be described symbolically (e.g., a number a lot greater than 2,030,498) and the other verbally (e.g., a number greater than fifteen million).

> **TEACHING TIP.** Some students are more comfortable with verbal descriptions of numbers. Others prefer symbolic descriptions.

> *Option 1:* A whole number is 60% of a whole number *n* that is less than 10. What could *n* be?
>
> *Option 2:* A whole number is 120% of a whole number *n* that is less than 10. What could *n* be?

Because the percent is a whole number, students must consider what possibilities there are. Most students are likely to select 60% of 5, which is 3, or 120% of 5, which is 6.

Regardless of the option selected, follow-up questions such as these could be asked:

- *How did you know that* n *was not 8?*
- *Was the percent more or less than* n? *How did you know?*
- *Why might it help to think of the percent as a fraction to solve the problem?*
- *What was* n? *What was the required percent of it?*

Variations. Instead of using 60% and 120%, 75% and 125% might be used instead so that students consider **multiples** of 4 instead of multiples of 5.

✴ **BIG IDEA.** **Classifying numbers provides information about the characteristics of those numbers.**

> **Option 1:** You factor a number into **primes**. There are two different prime factors. What else do you know about the number?
>
> **Option 2:** You factor a number into primes. There are two prime factors, but one is repeated. What else do you know about the number?

In both situations, students are asked to explore numbers of a certain type. In **Option 1**, students might recognize that numbers with two prime factors have four factors, have **factor trees** with two branches, and are called **composite numbers**.

In **Option 2**, students might recognize that the number has six factors, one of the factors is 1, the factor tree might have four branches, and the number is called composite. A possible factor tree for **Option 2** is:

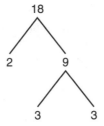

No matter which option students choose, follow-up questions such as these would be appropriate:

- *Could the number be prime? How do you know?*
- *Does the number have more than three factors? How do you know?*
- *How many factors does it have?*
- *How could you use a diagram to show that?*

> Draw the graph for your chosen option, where GCF means **greatest common factor**. What do you notice? What other graph would look a lot like the one you have drawn?
>
> *Option 1:* $y = GCF(5,x)$ *Option 2:* $y = GCF(6,x)$

The suggested graphs are unusual ones for students to consider, but they are useful to provide students with insight into the meaning of greatest common factor.

The difference between the graphs in *Options 1* and *2* is simply the complexity. Because 6 has factors, whereas 5 is prime, the graph for *Option 2* has more "hills":

Option 1:

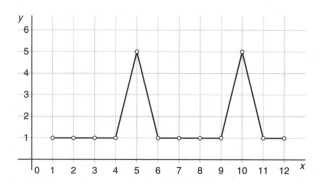

Option 2:

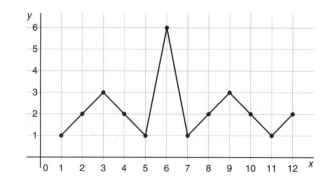

Similar graphs would be $y = GCF(prime,x)$ and $y = GCF(composite with 2 prime factors,x)$, for example, $y = GCF(3,x)$ and $y = GCF(15,x)$.

Relevant questions for both tasks include:

* *What is the greatest y value for your graph? The smallest? How could you have predicted those values?*
* *When does your graph reach its highest points? Why?*
* *When is your graph at its lowest points? Why?*
* *What second graph did you choose? Why did you choose that one?*
* *How can knowing the value for* p *in* y = GCF(p,x) *help you predict the shape of the graph?*

Variations. Students might consider the graphs of $y = LCM(p,x)$ (**least common multiple**) instead.

TEACHING TIP. Asking students to create a similar situation helps them to consider the innate structure of the situation they are exploring.

✳ **BIG IDEA.** There are many equivalent representations for a number or numerical relationship. Each representation may emphasize something different about that number or relationship.

> Choose one of the numbers below. Show at least four or five ways to represent it.
>
> **_Option 1:_** 0.0003   **_Option 2:_** 4,200,000,000

Students at this level are still making sense of very large numbers or decimals with many decimal places. This task allows students a choice about which of those types of numbers to explore.

Questions applicable to both tasks include:

- *Would it be easy to represent your number with* **base ten blocks**? *Why or why not?*
- *How could reading the number aloud help you come up with a representation for it?*
- *What kind of picture could you use to represent your number?*
- *Is your number big or small? What makes you call it big or small?*
- *If you wrote the number in words, how many words would you need to use?*
- *Can you write a number that is a very different size but requires the same number of words?*

Variations. Students might be given, instead, numbers in scientific notation, or decimals that are a little easier to represent (e.g., hundredths or thousandths).

✳ **BIG IDEA.** Numbers are compared in many ways. Sometimes they are compared to each other. Other times, they are compared to benchmark numbers.

> Draw a picture that would help someone compare the given pair of numbers.
>
> **_Option 1:_** $\sqrt{64}$ and $\sqrt{256}$ **_Option 2:_** $\sqrt{64}$ and $\sqrt{66}$

This task encourages students to compare numbers as well as represent them. In each instance, students are likely to use the notion that the square root of a number is the side length of a square with a particular area to help them with their picture. In **_Option 1_**, one square root is a whole number multiple of the other, so the square can simply be scaled up. In **_Option 2_**, the numbers are much closer together, so the relationship is quick to see, but drawing the square root might be more difficult because the square root, in one case, is not a whole number.

Regardless of the option selected, follow-up questions such as these could be asked:

- *How much is $\sqrt{64}$?*
- *Which square root is greater? How do you know? Is it a lot greater?*
- *Is the second square root a whole number or not? How do you know?*
- *What did your picture look like?*

Variations. Instead of using the specific values suggested, alternate values can be suggested, perhaps where neither is a simple square root.

> **_Option 1:_** A number is between 4π and $\sqrt{161}$. What might it be?
>
> **_Option 2:_** A number is between 4π and 5π. What might it be?

Both tasks require students to use the number 4π, but only one of them really requires students to estimate its value. In **_Option 2_**, a student might, for example, respond with the value 4.5π without actually estimating its value. In the case of **_Option 1_**, the two values are fairly close, so students will need to think carefully about the value to choose.

Relevant questions for both tasks include:

- *Did you need to calculate the value of 4π? Why or why not?*
- *Is your number more than 20? How do you know?*
- *How do you know your number is in the right range?*

Variations. An alternative to **_Option 2_** might be asking for a number between $\pi/2$ and π, because students might relate this to angle measurements.

✹ **BIG IDEA.** The operations of addition, subtraction, multiplication, and division hold the same fundamental meanings no matter the domain to which they are applied. Each operation has an inverse operation.

> **_Option 1:_** Write $\frac{2}{3}$ as the quotient of two other fractions. One must be greater than 2.
>
> **_Option 2:_** Write $\frac{2}{3}$ as the product of two other fractions. One must be greater than 2.

Each option requires students to consider products or quotients of fractions. By stating the condition that one number must be greater than 2, the problem is made slightly more challenging. The tasks are set up to make it clear to students that alternate answers can be obtained by either multiplying both values by the same amount (in the case of **Option 1**) or multiplying one term and dividing the other by the same amount (in the case of **Option 2**).

A possible solution to **Option 1** is $\frac{8}{3} \div 4$, but other solutions could involve dividing by a fraction, for example, $\frac{5}{3} \div \frac{5}{2}$. A possible solution to **Option 2** is $\frac{1}{12} \times 8$.

No matter which option students choose, follow-up questions such as these would be appropriate:

- *How did you solve the problem?*
- *Is there more than one possible solution? How do you know?*
- *Could both terms be greater than 2? Explain your thinking.*
- *Once you get one solution, what's an easy way to get another one?*

TEACHING TIP. Often changing the operation in a task results in a task that is parallel but has some fundamental differences. Exploring both operations at the same time helps students see what the differences between the operations really are.

✹ BIG IDEA. There are many algorithms for performing a given operation.

> **Option 1:** Describe two different ways to calculate 0.750 × 1.750.
>
> **Option 2:** Describe two different ways to calculate 0.750 ÷ 1.750.

Decimals that are easily described as fractions are deliberately chosen for this task to make it easy for students to consider alternate algorithms.

In **Option 1**, they might multiply $\frac{3}{4} \times \frac{7}{4}$, might subtract $1.750 \div 4$ from 1.750 or might multiply 0.750 by 2 and then subtract $0.750 \div 4$. In **Option 2**, they might multiply $\frac{3}{4} \times \frac{4}{7}$, might divide $\frac{3}{4}$ by $\frac{7}{4}$ using common denominators (seeing that the result is $\frac{3}{7}$), or might use an equivalent fraction (e.g., $\frac{75}{175}$).

Questions applicable to both tasks include:

- *Is the result more than or less than 1?*
- *Why does that make sense?*
- *Which of the ways you calculated the result did you think was easier? Why was it easier?*
- *Would you have used the same strategies if the values had been 0.85 and 1.850 instead? Explain your thinking.*

PARALLEL TASKS FOR GRADES 9–12

✸ **BIG IDEA.** **Numbers tell how many or how much.**

> **_Option 1:_** The **volume** of a cube is 29,000 cm³.
>   Estimate the side length.
>
> **_Option 2:_** The volume of a cube is 2,900 cm³.
>   Estimate the side length.

Students must estimate a **cube root** in either option. Some students might find **_Option 1_** simpler because they merely need to multiply the cube root of 29 by 10. More estimation might be required for **_Option 2_**.

Regardless of the option selected, follow-up questions such as these could be asked:

- *What operation do you perform to determine the side length? Why that operation?*
- *Is estimating your cube root as easy as estimating the cube root of 29? Why or why not?*
- *What estimate did you use? Why that one?*

Variations. Different values can be used for the volume, perhaps even a much smaller value to simplify the question sufficiently for students for whom cube roots might be too difficult. Alternatively, the areas of squares can be used rather than volumes of cubes.

> **_Option 1:_** A whole number power of $\frac{3}{5}$ is between $\frac{1}{10}$ and $\frac{1}{20}$. What is the power?
>
> **_Option 2:_** A power of $\frac{3}{5}$ is about 0.8. What is the power?

In each case, students deal with the notion that the effect of raising a fraction between 0 and 1 to a power is different from the effect of raising a whole number greater than 1 to a power. The fraction $\frac{3}{5}$ was selected to make it easier for students to use a decimal equivalent if they choose.

Students who pursue **_Option 1_** are likely to realize that $(\frac{3}{5})^2$ is 0.36 (i.e., less than the starting number of 0.6). To get a result that is even smaller—to fall within the specified range—they would probably continue to try higher powers.

Students who pursue **_Option 2_** need to realize that they must use a fractional power. This option may seem more difficult at first glance, because many students would not have considered using a fractional power.

Questions applicable to both tasks include:

- *What happens to $\frac{3}{5}$ when you square it?*
- *How does squaring $\frac{3}{5}$ help you figure out what to do?*
- *How do you know the power isn't 1?*

> ___Option 1:___ A negative integer is the **logarithm** (base 10) of the
> number A. What do you know about A?
>
> ___Option 2:___ A number of the form ☐.5, where ☐ is a whole number,
> is a logarithm (base 4) of the number A. What do you
> know about A?

In both circumstances, students need to know what logarithms are. Some students will be more comfortable with base-10 logarithms, although the fact that the logarithm is negative means that students must deal with fractions or decimals less than 1. Even though logarithms other than to base 10 might seem more difficult to students, they might notice that because 4 is a perfect square, a power like 42.5 is a whole number.

Relevant questions for both tasks include:

- *Could the number be greater than 10?*
- *Could the number be 0?*
- *Is the number more or less than the logarithm? Why?*

✳ **BIG IDEA.** **There are many equivalent representations for a number or numerical relationship. Each representation may emphasize something different about that number or relationship.**

> ___Option 1:___ Write the repeating decimal 0.234343434 . . . as a fraction.
>
> ___Option 2:___ Write the repeating decimal 0.010010010 . . . as a fraction.

Some students have significantly more difficulty writing repeating decimals as fractions when not all of the digits repeat. For this reason, ___Option 2___ is provided as an alternative.

There is only one solution for each task: for ___Option 1___, a fraction equivalent to $\frac{232}{990}$ $[\frac{2}{10} + (\frac{1}{10} \times \frac{34}{99})]$, and for ___Option 2___, and a fraction equivalent to $\frac{10}{999}$.

Relevant questions for both tasks include:

- *What feature of the decimal indicated to you that it could be written as a fraction?*
- *How might you have estimated the value of the fraction before calculating it?*
- *Suppose the tenths digit had been 1. How would your fraction change? How do you know?*

Complete the task without using a calculator.

Option 1: Simplify $\sqrt{11{,}250}$.

Option 2: Simplify $\sqrt{72} + 7\sqrt{162} + \dfrac{32}{\sqrt{8}}$

Both tasks require students to simplify **radicals**. More simplification is required in **_Option 2_**, including **rationalizing** of the denominator.

No matter which option students choose, follow-up questions such as these would be appropriate:

- *Is the value of the expression closer to 100 or 200? How do you know?*
- *Was the simplified answer a rational number? Why or why not?*
- *Were you able to simplify to a **simple radical**? Why or why not?*

Option 1:   The answer is 5^{20}. What might the question be?

Option 2:   The answer is $4^{\frac{2}{3}}$. What might the question be?

Students could use either definitions of whole number or rational exponents or exponent laws to answer either of these tasks. For example, for **_Option 1_**, a student might say, *What do you write if you want someone to multiply 5 by itself 20 times?* but might also ask, *What is $(5^4)^5$ or $5^{17} \times 5^3$?* For **_Option 2_**, a student might ask for the cube root of 16 or for the square of the cube root of 4, or might ask, *What is $4^{\frac{1}{6}} \times 4^{\frac{1}{2}}$?*

Questions applicable to both tasks include:

- *What are the **base** and exponent in your expression? What role does each play?*
- *Is it easier to write your power as a product or a sum? Explain.*
- *Is it easier to write your power as a product or a quotient? Explain.*
- *Is it easy to estimate the value of your power?*
- *How did you represent your power in different ways?*

TEACHING TIP. Some students might find work with fractional exponents challenging. A task involving a whole number exponent is a suitable parallel.

✳ **BIG IDEA.** Numbers are compared in many ways. Sometimes they are compared to each other. Other times, they are compared to benchmark numbers.

> Order the given set of numbers from least to greatest. Do not use a calculator.
>
> **_Option 1:_** $\sqrt{200}$, 2.5^3, 4^2, $\sqrt{260}$ **_Option 2:_** $\frac{1}{2}^4$, $4^{\frac{1}{2}}$, $3^{\frac{1}{3}}$, 1^8

This question requires students to estimate powers or roots. In **_Option 1_**, students should immediately recognize that $\sqrt{200} < \sqrt{260}$, but they will have to do more work with the other values. All values are actually close to 15. In **_Option 2_**, students should immediately recognize that powers of 1 are 1 and (one hopes) that $4^{\frac{1}{2}}$ is 2. In this case, all values are between 1 and 2, inclusive.

Regardless of the option selected, follow-up questions such as these could be asked:

- *Which values did you find easiest to calculate or estimate first? Why?*
- *Some of the values were close. How did you handle figuring out which was the greatest and which was the least?*
- *How did you estimate the values you had to deal with?*

✳ **BIG IDEA.** The operations of addition, subtraction, multiplication, and division hold the same fundamental meanings no matter the domain to which they are applied. Each operation has an inverse operation.

> What is the middle entry in the given **matrix**? How do you know?
>
> **_Option 1:_**
>
> $$\begin{bmatrix} 3 & 9 \\ 0 & 1 \\ 2 & 8 \end{bmatrix} \times \begin{bmatrix} -1 & 3 & 9 \\ 0 & 2 & 4 \end{bmatrix}$$
>
> **_Option 2:_**
>
> $$4 \times \begin{bmatrix} 3 & 9 & 0 \\ 0 & 1 & -1 \\ 3 & -2 & -4 \end{bmatrix}$$

Matrix multiplication is quite a bit more complex than **scalar multiplication**; therefore, students who find matrix multiplication difficult might select **_Option 2_**. By asking for the middle entry, the teacher can observe whether students simply mechanically calculate all entries or whether they focus on the required entry only.

Relevant questions for both tasks include:

- *If you completed the multiplication, how many rows and columns would your matrix have? How do you know?*
- *Can you tell before you multiply whether there will be zero elements in the matrix? If there are any, how many?*
- *How did you calculate the middle entry?*

SUMMING UP

MY OWN QUESTIONS AND TASKS

Lesson Goal: **Grade Level:** _____

Standard(s) Addressed:

Underlying Big Idea(s):

Open Question(s):

Parallel Tasks:
Option 1:

Option 2:

Principles to Keep in Mind:
- All open questions must allow for correct responses at a variety of levels.
- Parallel tasks need to be created with variations that allow struggling students to be successful and proficient students to be challenged.
- Questions and tasks should be constructed in such a way that will allow all students to participate together in follow-up discussions.

The six big ideas that underpin work in Number and Operations were explored in this chapter through 45 examples of open questions and parallel tasks, as well as variations of them. The instructional examples provided were designed to support differentiated instruction for students at different developmental levels, targeting two separate grade bands: grades 6–8 and grades 9–12.

Although Number and Operations becomes a less prominent strand in and of itself at the high school level, a strong foundation in number and operations facilitates success in all of the other strands. It is critical to help those who are underachieving in number and operations to meet more success and develop confidence. A form such as the one shown here can serve as a convenient template for creating your own open questions and parallel tasks. The Appendix includes a full-size blank form and tips for using it to design customized teaching materials.

Geometry

DIFFERENTIATED LEARNING ACTIVITIES in geometry are derived from applying the NCTM process standards of problem solving, reasoning and proof, communicating, connecting, and representing to content goals of the NCTM Geometry Standard, including

- analyzing characteristics and properties of two-dimensional (2-D) and three-dimensional (3-D) shapes, and developing mathematical arguments about geometric relationships
- specifying locations and describing spatial relationships
- applying transformations and using symmetry to analyze mathematical situations
- using visualization, spatial reasoning, and geometric modeling to solve problems (NCTM, 2000)

TOPICS

Before beginning the task of differentiating student learning in geometry, it is useful for teachers to have a good sense of how the topics in the strand develop over the grade bands. The NCTM *Curriculum Focal Points* (NCTM, 2006), which suggest what mathematical content should be the focus at each grade level through grade 8, were used as the basis for recommendations made in this resource for grades 6–8; the NCTM *Principles and Standards for School Mathematics* (NCTM, 2000) helped form the basis for the material for the higher grades. For a teacher at a particular grade level, it can be helpful to be aware of where students' learning is situated in relation to what learning has preceded the present grade band and what will follow.

Some topics that are considered geometry in one jurisdiction might be regarded as measurement topics in another jurisdiction. For example, **trigonometry** is focused on measurement but is built on the geometric property of **similarity**. The Glossary and the Index of Big Ideas at the back of this resource can help the reader discover where particular topics that might be in either domain are covered.

Grades 6–8

Within this grade band, students focus on **decomposing** shapes in two dimensions and three dimensions to support work in area and volume, although they might also explore composing and decomposing shapes in the context of **tessellations**. They also solve more complex geometry problems than younger students would, including locating shapes on coordinate grids and using the Pythagorean theorem to relate lengths of sides in a right triangle.

Another topic at this level is similarity, the result of enlarging or reducing a shape so that the overall proportions, but not necessarily the size, remain the same.

Many students also construct geometric objects using a variety of tools, considering the properties of those objects to facilitate the constructions.

Grade 9–12

Within this grade band, students spend a significant amount of time using coordinate systems to study and develop properties of shapes; consider necessary and sufficient conditions for congruence and similarity; develop the concept of proof, using it to investigate a variety of geometric **conjectures**; work with 3-D objects such as **planes**, **polyhedra**, and **vectors** (in both 2-D and 3-D); study and apply **circle properties**; and sometimes study issues of **connectivity**, as in **networks** and graphs. In some jurisdictions, students also study **non-Euclidean geometry** (e.g., the geometry of the sphere).

THE BIG IDEAS FOR GEOMETRY

Coherent curricula in geometry that meet NCTM content and process standards (NCTM, 2000) and support differentiated instruction can be structured around the following big ideas:

- Both quantitative and qualitative attributes of a geometric object can impact measurements associated with that object.
- Many properties and attributes that apply to 2-D shapes also apply to 3-D shapes.
- How a shape can be composed and decomposed, or its relationship to other shapes, provides insight into the properties of the shape.
- There are many representations of a geometric object or a relationship between geometric objects.
- Locations of objects can be described in a variety of ways.

The tasks set out and the questions asked while teaching geometry should be developed to evoke these ideas. The following sections present numerous examples of application of open questions and parallel tasks in development of differentiated instruction in these big ideas across two grade bands.

OPEN QUESTIONS FOR GRADES 6–8

> **OPEN QUESTIONS** are broad-based questions that invite meaningful responses from students at many developmental levels.

✴ **BIG IDEA.** Both quantitative and qualitative attributes of a geometric object can impact measurements associated with that object.

> A 3-D shape has 10 vertices. What could the shape be?

To respond to the question, students might need to be reminded that vertices are corners of the shape. A **vertex** can be thought of either as the point where two **edges** meet or where three or more faces meet.

For students who struggle to come up with a response, questions that could be asked include:

- *Could the shape be a square* **prism**? *Why or why not?*
- *If it were a prism, how many edges would each* **base** *have? Why?*
- *Does the shape have to be a prism?*

Variations. Rather than 10 vertices, the number could be some other even number (to allow for either a prism or **pyramid** response). Alternatively, the number of edges or faces could be provided instead.

> A shape only has one kind of symmetry. What could the shape be?

At this level, students have learned about both **reflection symmetry** and **turn symmetry**. This question allows a student to consider both types, in either two dimensions or three dimensions. One student might choose a 2-D shape with reflection symmetry but no turn symmetry, another might think of a 2-D shape with turn symmetry but no reflection symmetry, and a third student might select a 3-D shape with reflection symmetry but no turn symmetry.

Variations. Instead of asking for a shape with one kind of symmetry, a student might be asked to describe a shape with both types of symmetry, or a shape with no symmetry.

> Two triangles are almost congruent. Describe the triangles as completely as you can. Tell why they are almost congruent.

Students might focus on side measurements, angle measurements, or some combination thereof to describe triangles that are almost congruent. It would be interesting to see how close they feel the measurements need to be for the triangles to be considered almost congruent.

Variations. Instead of expecting the two triangles to be almost congruent, the shapes could be almost similar.

TEACHING TIP. Using the word "almost" in a question opens up the possibility for more diverse responses than might be prompted by a more specific question.

✳ **BIG IDEA.** **Many properties and attributes that apply to 2-D shapes also apply to 3-D shapes.**

> A 3-D shape shares many of the properties of a 2-D rectangle. What might the shape be? Why?

To respond to this question, students need first to decide what properties of a rectangle upon which they will focus. Then they need to consider which 3-D shapes might share those properties. It would be useful for the teacher to display both a rectangle and a number of 3-D models to support students as they respond to this question.

If students have difficulty getting started, the teacher could provide scaffolding with questions such as these:

- *What is special about a rectangle?*
- *Could a round 3-D shape be like a rectangle or not?*
- *Is there more than one possible 3-D shape that is like a rectangle?*

Variations. Instead of sharing properties with a rectangle, the 3-D shape might share properties with a square, a circle, or an **isosceles triangle**.

TEACHING TIP. Instead of specifying how many properties have to be shared among objects being compared, a question is more open if it simply indicates that "many" properties are shared. Using descriptors such as "many," "few," or "some" removes some of the pressure on students that would result if a specific number were used instead.

> A shape has a lot of **diagonals**. What might the shape be? How many diagonals does it have?

Students need to recognize that diagonals join vertices of a shape that are not already connected by edges. Some students might choose simple 2-D shapes (e.g., **quadrilaterals**), others might name more complicated 2-D shapes, and still others might select 3-D shapes.

The important mathematical discussion prompted by this question revolves around the counting of the diagonals. Some students might look for relationships that would allow them to predict the number of diagonals by knowing the number of vertices of the shape, but other students might consider whether different types of shapes have more or fewer diagonals than other types of shapes. For example, a cube has fewer internal diagonals than an **octagon** has, but that is not the case if diagonals on faces are also included. This will generate an interesting discussion.

> How is figuring out how this structure could be cut into congruent halves like figuring out how to cut a square into congruent halves? How is it different?
>
>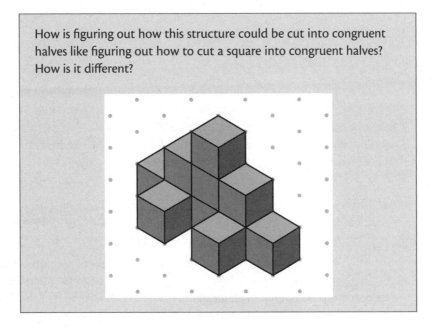

To respond to this question, it may be necessary to remind students that shapes that are congruent are the same size and shape, that is, identical other than in position. Some students might notice that only a particular vertical cut can be used to cut the given structure into congruent halves, although there are many more ways to cut a square into congruent halves.

✳ **BIG IDEA.** How a shape can be composed and decomposed, or its relationship to other shapes, provides insight into the properties of the shape.

> Choose one of the **tangram** shapes. Then make another shape congruent to that one by combining other pieces.

A tangram is a set of seven specific shapes that can be combined to form a square. Typically, students are told what shape they must create with the smaller shapes. Providing the student with a choice opens up the question.

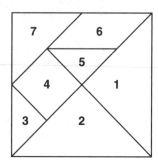

For example, a student might create square 4 using triangles 3 and 5; another student might create triangle 1 using shapes 6, 3, and 5 or shapes 7, 3, and 5.

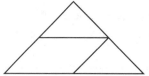

Variations. A tangram diagram could be provided instead of cut-out shapes. This would force students to use visualization skills in a more sophisticated way, but is a bit less accessible than allowing students to use cut-outs.

> What shapes can you create by combining two congruent **hexagons**?

This open question encourages students to experiment with the two hexagons. Although some students will only consider **regular** hexagons, others will consider more varied possibilities. Whereas some students will assume that the shapes must share an edge or vertex and not overlap, other students might choose to overlap the shapes. For example:

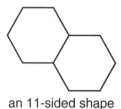

an 11-sided shape

a 12-sided design

a bigger hexagon

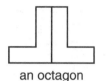

an octagon

For students who do not know how to begin, questions such as these could be helpful:

- *Could a hexagon have all **right angles**?*
- *Could it have all **acute angles**?*
- *How might two hexagons touch?*
- *How would you arrange the two hexagons to get a shape with lots of sides? With few sides?*

Variations. Rather than combining two hexagons, students could be asked to combine two other shapes.

> You combine a shape with a **surface area** of 22 cm² with a shape with a surface area of 24 cm². What could the surface area of the combined shape be?

Although there is a measurement element to this question, the focus is really on the geometric arrangement of two shapes. The question is set up in such a way that a student might simply consider two simple shapes that meet the criteria, for example, a $1 \times 2 \times 3$ rectangular prism and a $2 \times 2 \times 2$ cube. The student could then consider different ways to combine the shapes to see how surface area is lost. For example, the two shapes might simply be stacked (combination at left below, with a surface area of 38 cm²), or one might be embedded within the other (combination at right below, with a surface area of 36 cm²).

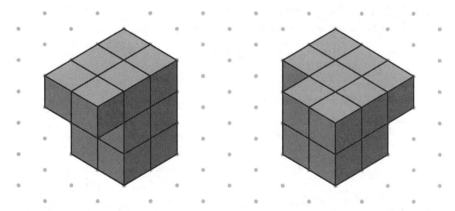

Variations. The shapes being combined could have different surface areas. Alternatively, three shapes could be suggested, or volumes could be provided rather than surface areas.

A shape is $\frac{1}{8}$ of an octagon. What might the shape be? Is there more than one possibility?

Many students will assume that the octagon has to be a regular octagon and that $\frac{1}{8}$ of it is a triangle with one side of the octagon as a base and the center of the octagon as the third vertex of the triangle.

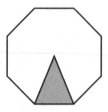

By asking students for another possibility, they are more likely to think of other sorts of octagons, such as the ones below:

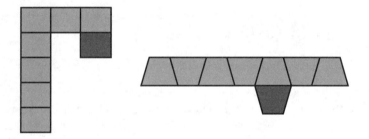

Variations. Instead of using $\frac{1}{8}$ of an octagon, a different shape and a different fraction could be used. Alternatively, students could be given a shape and told it is a particular fraction of another shape.

✳ BIG IDEA. There are many representations of a geometric object or a relationship between geometric objects.

What information does this **base plan** provide?

What information does it not provide?

3	3
3	3
1	1

Front

This question is only accessible once students know what a base plan is. The base plan shown on page 96 is a description of the structure at the right:

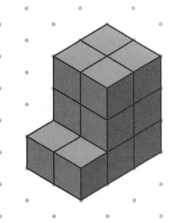

The base plan shows how high each part of the structure is. However, students might suggest many different things the base plan does or does not show. It shows, for example, that the base has 6 sections. It shows that the maximum height is 3 cubes. It shows that there is some symmetry. It does not indicate the color of the cubes that were used to build the structure. It does not show how big the structure is (i.e., what size cubes are used).

The teacher can follow up by asking students whether the base plan feels, to them, more connected to a **net** describing the structure, or an **isometric drawing**, or an **orthographic** (front, side, top **views**) representation.

> The net of a shape has 10 separate parts. What might the shape look like?

By not showing the net and by using the fairly vague phrase "separate parts," students have greater latitude in answering this question. A student might decide that the net is made up of 9 triangles and a 9-sided base (representing a pyramid), such like the one below:

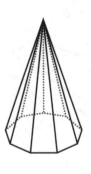

Another student might decide the shape is an octagonal prism:

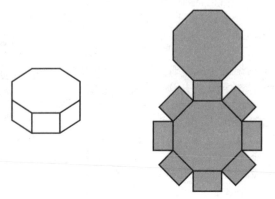

A third student might take a different approach altogether, and let the "parts" of the net represent units of the surface area of a $2 \times 1 \times 1$ rectangular prism:

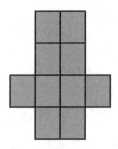

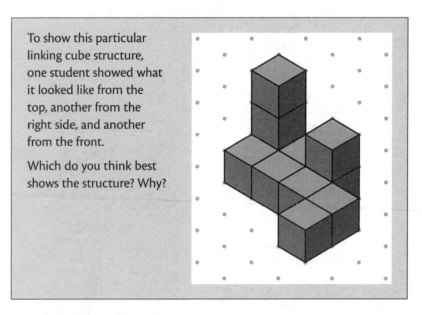

To show this particular linking cube structure, one student showed what it looked like from the top, another from the right side, and another from the front.

Which do you think best shows the structure? Why?

This question focuses students, indirectly, on the value of having all three views of a 3-D object. In deciding which view is best, students need to think about what is lacking from any particular view.

If students struggle to come up with a response, the teacher could ask scaffolding questions such as:

- *What part of the structure would the front (or side, or right) view show?*
- *What part does it not show? Why does it not show that part?*

TEACHING TIP. Asking students for their opinions on a matter is a valuable strategy to take the risk out of responding to a question.

✴ **BIG IDEA.** Locations of objects can be described in a variety of ways.

> An isosceles triangle has one vertex at (1,3). Where might the other vertices be? What else do you know about the triangle you have described?

Students will have to decide whether (1,3) is the **apex** of the isosceles triangle or whether the point (1,3) is on one of the equal sides. They will then need to think about how to describe the triangle. Some possible triangles are shown here:

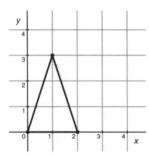 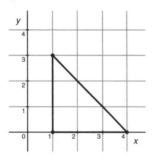

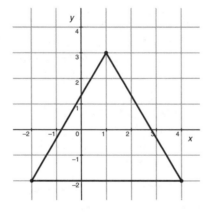

The students might choose to describe their triangles by listing the other vertices or they might choose to use terms such as *acute, right,* or *obtuse.*

OPEN QUESTIONS FOR GRADES 9–12

✳ **BIG IDEA.** Both quantitative and qualitative attributes of a geometric object can impact measurements associated with that object.

> Andrew said that deciding whether lines are parallel has nothing to do with measurements, but Felix says it does. With whom do you agree? Why?

In this situation, asking students with whom they agree is a fairly safe entry point, particularly because a case can be made for either answer. Any student can begin by taking a position; then he or she can work with other students who have taken that position to help develop a good argument.

Some students will focus on the fact that parallelism is defined in terms of whether lines intersect, not in terms of measurements; others will realize that measurements can be used to determine whether two lines intersect or not.

For students having difficulty getting started, questions that could be asked include:

- *What makes two lines parallel?*
- *Is there another way to tell?*
- *What kinds of measurements—length, area, volume, etc.—are possible to consider? Why these?*

Variations. Rather than asking about parallelism, a teacher could ask about perpendicularity.

> You know that one angle in a triangle is 53°. What else do you know about the triangle?

This question is accessible because there are many ways to approach it and there are many answers. Some students might simply draw a 53° angle and complete the shape to form a triangle. They might provide the side lengths of their triangles, describe the triangles as either **acute** or **obtuse** or right, describe them as isosceles or **scalene**, or perhaps describe the other angles. Other students will use what they know about the sum of the angles of a triangle and indicate that the other two angles total 127° in measure.

If students do not know how to begin, the teacher could provide scaffolding with questions such as these:

- *What makes one triangle different from another one?*
- *What attributes of a triangle can you measure?*

- *What words can you use to describe triangles?*
- *Could the other angles of the triangle both be less than this one?*
- *Can you be sure whether the 53° angle is part of the longest side of the triangle or not? Explain.*

TEACHING TIP. By not indicating what must be said about a shape, but simply allowing students to say what they wish, a question becomes much more accessible.

> Two of the vertices of a quadrilateral are (1,3) and (5,3). The other vertices are also in Quadrant 1. What kind of quadrilateral could it be? Explain.

By choosing two of the vertices in a horizontal line, the question is made more accessible for some students. Students could easily create a simple rectangle in Quadrant 1. However, other students could be challenged to make more complicated shapes—for example, **concave** quadrilaterals, **trapezoids**, **kites**, or parallelograms. Students will have to use properties of the shapes to describe what sort of quadrilateral they have created and how they know that the vertices used create that shape.

Variations. Rather than asking for quadrilaterals, students might be asked to create other regular shapes. Alternatively, different pairs of vertices could be provided or the quadrant(s) for the other vertices could be varied.

TEACHING TIP. Sometimes it is wise to ask a question that could be answered fairly simply, but privately coax students who need a challenge to look for a more sophisticated answer. This is in contrast to asking a more challenging question that leaves out students not up to the challenge.

> The diagonals of a quadrilateral meet at a 90° angle. How could you decide which quadrilateral it might be?

Some students may know that if the diagonals of a quadrilateral meet at a right angle, then the quadrilateral is either a **rhombus** or a kite. Students might use a ruler to measure the sides of the shape. If all four sides are equal, the shape is a rhombus; if not, and pairs of adjacent sides are equal, the shape is a kite.

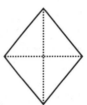 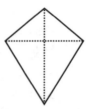

Alternatively, students might realize that the diagonals **bisect** each other in a rhombus, but not necessarily in a kite. They could either measure or fold along the diagonals to check the relative lengths of the two parts of each diagonal. Still other students, who may not use prior knowledge about diagonals of kites or rhombi, might simply draw two line segments meeting at a 90° angle and use the endpoints as vertices. They could examine the shape created to decide what sort of quadrilateral it is.

How could you draw a right angle without using a protractor?

This question builds on different types of geometric knowledge. Some students will use concepts of symmetry. They can fold a piece of paper horizontally and then vertically, creating two lines of symmetry meeting at a right angle. Other students will use circle properties. After drawing a circle and a diameter, they can connect the ends of the diameter to another point on the circumference of the circle to create a right angle. Yet other students might draw a rhombus or kite, and then draw the diagonals to form a right angle.

For students who struggle to come up with a response, questions such as these could be helpful:

- *Suppose you were thinking of **polygons**—where would you see right angles?*
- *Suppose you were thinking of circles—where would you see right angles?*

Think of several pairs of measurements related to a circle that are equal. How do you know they are equal?

Students might consider the fact that all radii or all diameters of a circle are equal in length to get a fairly simple solution. They can use the definition of a circle to describe why the radii are equal and the relationship between radii and diameters to explain why diameters are equal.

However, students might also recognize that **chords** of a circle an equal distance from the center are equal in length or that the two sections of a chord split by the perpendicular to the center of the circle are also equal.

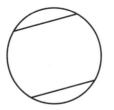 

What makes an **ellipse** special?

There are many possible answers to this question. Some students will describe the algebraic form of an ellipse,

$$\frac{x^2}{a^2} + \frac{y^2}{b^2} = 1,$$

indicating that this particular form only describes an ellipse and that is what makes the ellipse special.

Some students might suggest that what makes an ellipse special is that it is like a circle, but not quite a circle. Further probing could get them to explain how it is like a circle. Others might talk about places in the real world where ellipses occur, e.g., elliptical orbits.

Some students might simply refer to the fact that the ellipse is special because it has two lines of symmetry as well as turn symmetry. Still other students will see the ellipse as special because it is a **cross-section** of a cone and there are not that many shapes that are cross-sections of cones.

Variations. Rather than asking about an ellipse, one might ask why a circle, a parabola, a **hyperbola**, or maybe a **bell curve**, is special.

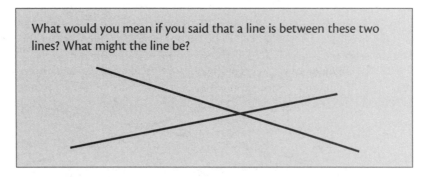

What would you mean if you said that a line is between these two lines? What might the line be?

Students are accustomed to being asked whether lines intersect or not, but have probably not had experience talking about lines between other lines. It is exactly this lack of familiarity that makes the question open—students feel freer to make their own interpretation of what "between" means in this context.

Some students are likely to assume that the third line must be the line that cuts the narrow **vertical angles** in half.

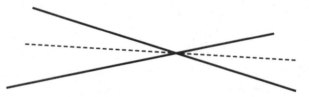

Other students may assume that any line that stays between the two lines, horizontally, is correct.

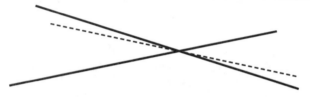

But some students might notice that the line could go through the point of intersection more vertically. For example:

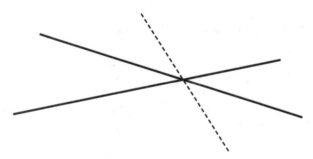

Some students might think entirely differently and draw a line segment between the two lines. They will need to be encouraged to decide whether the line of which the segment is part is actually between the two lines or not.

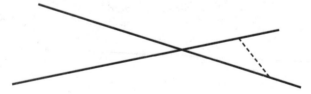

Interesting questions arising out of the original task become what slopes are possible for the "between" line, whether the line can be parallel to either original line, and so forth.

TEACHING TIP. Although we often shy away from ill-defined terms in mathematics, it is valuable to give students a chance to define those ill-defined terms to expose their thinking.

✴ **BIG IDEA.** Many properties and attributes that apply to 2-D shapes also apply to 3-D shapes.

> What do you think circles and spheres have in common?

This question offers a great deal of latitude. Students might focus on the symmetry of both shapes, the roundness of both shapes, the similarity of the algebraic representations on a coordinate grid for both shapes, the equidistant component of the definitions for both shapes, the fact that both are **limits** of a class of shapes (whether regular polygons or regular polyhedra), and so on.

If students have difficulty getting started, the teacher could ask scaffolding questions such as:

- *What makes a circle special?*
- *Does that same idea apply to a sphere?*
- *What attributes does a circle have?*
- *Does a sphere have those attributes?*

Variations. Instead of looking at circles and spheres, students might consider squares and cubes, or **equilateral triangles** and **tetrahedra**.

✴ **BIG IDEA.** How a shape can be composed and decomposed, or its relationship to other shapes, provides insight into the properties of the shape.

> You divide the parallelogram with vertices (1,0), (8,0), (2,3), and (9,3) by a vertical line. The left trapezoid has a lot more area than the right trapezoid. What vertical line might you have used? Explain.

This question assumes that students are familiar with Cartesian coordinates. Although there is not a requirement to provide the equation of the line (simply a request to describe it), many students will use their knowledge of equations of lines to respond to the question. Other students will describe how far the line is from points on the perimeter of the parallelogram.

The phrase "a lot more" is deliberately vague in order to provide more maneuvering room for students to respond. Two possible responses are shown on the next page.

The line is the one that goes up from the point (7,0).

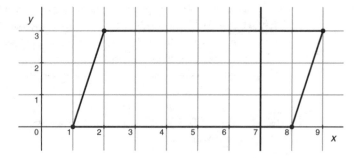

The line has the equation x = 6.

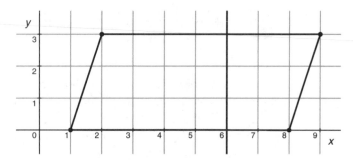

Variations. Rather than demanding a vertical line, students could be allowed to divide the parallelogram into trapezoids with a different relationship. Alternatively, a different initial shape could be used and the area relationship between the two parts could be changed: for example, they could be expected to be close but not equal in area.

✹ **BIG IDEA.** **There are many representations of a geometric object or a relationship between geometric objects.**

> You are trying to describe a specific plane in three dimensions. How might you do that?

This question is designed to allow students to recognize that there are a number of ways to describe a plane, whether an algebraic equation involving the variables *x*, *y*, and *z*; the coordinates of three points on the plane; a **linear combination** of two given vectors in 3-space; or a point and a vector perpendicular to the plane at that point.

> How is an isometric drawing of a cube with edge length 3 similar to one for a cube with edge length 4. How is it different?

Students could choose to focus on many different aspects of the isometric drawings to compare the two required drawings. For example, one student might talk about how to start with one of the drawings to change it to get to the other. Others might focus on the number of small cubes visible in each drawing. Still others might focus on measurement relationships in the drawing.

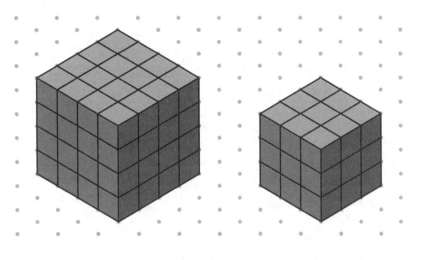

TEACHING TIP. Asking how two mathematical objects are alike and/or different is a useful strategy for creating open questions.

✹ **BIG IDEA.** Locations of objects can be described in a variety of ways.

> Mark wants to use measurements to describe the position of a spot on the ceiling in a room. What might he do?

The purpose of this question is to elicit the notion that while we need three coordinates to describe a position in space, we need only two coordinates if we know the plane upon which the point is located.

One student might make a cogent argument that three measurements are needed, including the height of the ceiling, whereas another student will take the ceiling height for granted. Even that student could describe the measures needed in different ways. For example, it might be a **polar coordinate** description rather than a Cartesian coordinate description that is used.

For students who do not know how to begin, questions that could be asked include:

- *Could you use a single measurement to describe a point on your desk? On a number line?*
- *Would you need more or fewer measurements to describe the location of the point if it were on an edge of the ceiling than somewhere near the middle?*
- *If I mark this point on a piece of paper, would you use length or area measurements to describe its location? Why?*

Tell about a shape that you think would be easier to describe using polar coordinates than Cartesian coordinates. Explain your thinking.

Students would need to understand polar coordinates to answer this question. But once they do, there are a number of simple solutions, as well as more complex ones. For example, a student might suggest that the equation $r = 4$ is simpler than the equation $x^2 + y^2 = 16$ because there is a lot less to write. Similarly, a student might suggest that the equation $\theta = 60°$ is a lot simpler than the equation $y = \frac{\sqrt{3}}{2}x$. Students familiar with more complicated curves, such as a spiral or a rose, will realize that the polar coordinate description is also much simpler than the Cartesian coordinate description.

Variations. The question could be reversed, asking for shapes that are easier to describe using Cartesian coordinates.

Describe different ways to tell how to get from point A to point B.

A ●

● **B**

Students familiar with **bearings** will know that one way to describe direction is by using bearings. Other ways involve the use of coordinates, whether Cartesian or polar, or vectors. Allowing students to choose how to describe the motion rather than asking them to do it in a certain way (e.g., with bearings) makes the question more open-ended and more accessible to a broad range of students.

Variations. A similar problem could be posed in 3-space, or by giving students the coordinates or A and B, although the latter is more likely to limit the range of responses students will give.

PARALLEL TASKS FOR GRADES 6–8

PARALLEL TASKS are sets of two or more related tasks that explore the same big idea but are designed to suit the needs of students at different developmental levels. The tasks are similar enough in context that all students can participate fully in a single follow-up discussion.

✸ ██ BIG IDEA. ██ **Both quantitative and qualitative attributes of a geometric object can impact measurements associated with that object.**

> _Option 1:_   A triangle has one 120° angle. What type of triangle might it be? What type of triangle can it not be?
>
> _Option 2:_   A triangle has one 60° angle. What type of triangle might it be? What type of triangle can it not be?

Some students might be attracted to _Option 1_ because they can be certain that the triangle could be labeled obtuse and cannot be equilateral. But they would need to apply additional reasoning to describe why it still might be isosceles (if the two other angles are 30°) or scalene. Students who choose _Option 2_ might recognize that the triangle could be equilateral or isosceles or scalene, or that it might be right, acute, or obtuse. It will be harder for them to indicate a triangle type it cannot be.

No matter which option students choose, follow-up questions such as these would be appropriate:

- _What types of triangles are there?_
- _Are there other types?_
- _What makes a triangle acute (or obtuse, or isosceles, etc.)?_
- _Can a triangle be more than one type of triangle?_

Variations. Students could be given the measures of one or two angles in a quadrilateral and asked a similar question.

TEACHING TIP. It is often a good idea to ask about what something might be but also what it might not be.

> _Option 1:_ Use a **pentagon** to tessellate the plane.
>
> _Option 2:_ Use a quadrilateral that is not a rectangle to tessellate the plane.

Students who choose _**Option 1**_ must realize that a regular pentagon cannot be used. However, there are a number of **irregular** polygons that can be used to tessellate a plane. An example is:

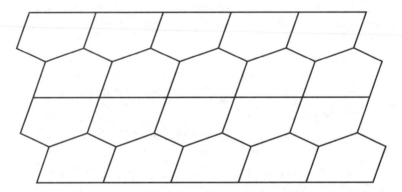

Students who choose _**Option 2**_, which might be simpler for many students, could choose a parallelogram or a **right trapezoid**, fairly simple shapes, or they might choose a more challenging quadrilateral. Examples are shown below:

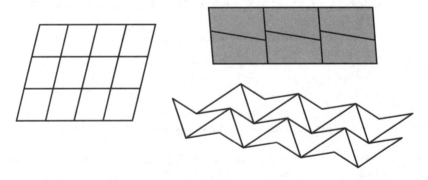

Questions applicable to both tasks include:

- _How many copies of your shape would you need to use before you were pretty sure it would tessellate the whole plane?_
- _Look at a point on your tessellation where shapes meet. How many angles meet at that point? Why does that make sense?_
- _What do you notice about shape edges that touch in your tessellation?_
- _Did you need to measure the angles in your shape to create the tessellation?_

Variations. Students might be asked to use other shapes with which to tessellate, for example, pentominoes (shapes made up of five touching unit squares).

> *Option 1:* The point (9,–2) is the top right vertex of a parallelogram.
>   What might the coordinates of the other vertices be?
>
> *Option 2:* The points (9,–2) and (1,–1) are two vertices of a
>   parallelogram. What might the coordinates of the other
>   vertices be?

Students must recognize that the opposite sides of a parallelogram must be both parallel and equal in length. Although students may assume vertical or horizontal orientations of one pair of sides of the parallelograms, this is not actually required, so there are many possible solutions in either case. Clearly, *Option 1* allows more flexibility (because only one vertex is designated), but some students might find *Option 2* simpler because two vertices are provided.

Regardless of the option selected, follow-up questions such as these could be asked:

- *In what quadrant(s) is your parallelogram?*
- *How did you ensure that the shape you created was a parallelogram?*
- *What did you do first, then next, to create the parallelogram?*
- *How could you use your coordinates to test that the shape you created is, in fact, a parallelogram?*

Variations. Students might be provided one or two sets of coordinates of other shapes instead (e.g., trapezoids or kites).

TEACHING TIP. One of the ways to create a set of parallel tasks is to provide either more or less information about a similar situation in one of the options.

✳ BIG IDEA. **Many properties and attributes that apply to 2-D shapes also apply to 3-D shapes.**

> *Option 1:* Create a nonstandard 2-D shape that has turn symmetry.
>
> *Option 2:* Create a nonstandard 3-D shape that has turn symmetry.

Students must recognize the difference between reflection and turn symmetry to approach this question. Although many students might prefer to use a 2-D shape, which is simpler to consider, other students might enjoy the challenge of a 3-D shape.

The question asks for a nonstandard shape so that students do not simply repeat what they have already learned about, for example, a square or a cube having

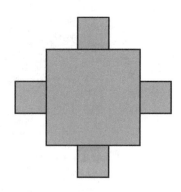

turn symmetry. Many students will likely start with a standard shape and vary it slightly to create the nonstandard shape. For example, a shape such as the one shown at the right would be a 2-D possibility:

Similarly, students could place small cubes centered on all of the faces of a larger cube to create a 3-D shape with turn symmetry.

Relevant questions for both tasks include:

* *How would you describe your shape?*
* *How do you know it has turn symmetry?*
* *What order of turn symmetry does it have?*
* *How did you come up with your shape?*

TEACHING TIP. Asking students to consider a nonstandard situation may make some students nervous initially, but ultimately is liberating in allowing them to develop their own lines of thinking.

✳ **BIG IDEA.** How a shape can be composed and decomposed, or its relationship to other shapes, provides insight into the properties of the shape.

> *Option 1:* How can dividing the red trapezoid pattern block into two shapes show that the red shape is $\frac{3}{2}$ of the area of the blue rhombus pattern block?
>
> *Option 2:* How can dividing the trapezoid below into three rectangles and two triangles help explain the formula for its area?
>
>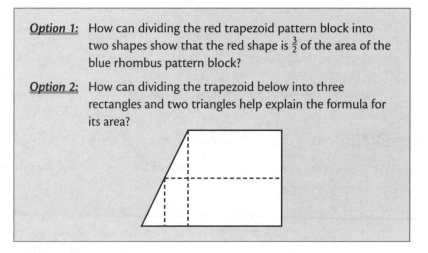

Each option helps students to see how areas of 2-D shapes can be calculated by decomposing and sometimes recomposing shapes.

In **Option 1**, the student could realize that because the blue shape is made of two triangles and the red shape is made of three of the same triangles, the red shape's area is $\frac{3}{2}$ of the blue shape's. Alternately, students might see that one can cut the trapezoid into a blue shape and another half blue shape, thus it is $\frac{3}{2}$ of the blue shape.

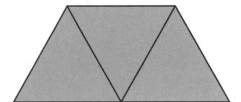 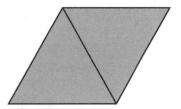

For **_Option 2_**, students might realize that the lower triangle could be rotated to complete a rectangle, and then relate the trapezoid area to the area of the rectangle made up of the same pieces. That rectangle has the same height as the original one and has a base that is the average of the original two bases.

No matter which option students choose, follow-up questions such as these would be appropriate:

- *If you cut up a shape and rearrange the pieces, does the total area change?*
- *Why might it be useful to cut up a shape to determine its area?*
- *How else could you cut up your shape to make it easy to relate the area of your shape to the area of another shape with an easy-to-calculate area?*

✳ **BIG IDEA.** **There are many representations of a geometric object or a relationship between geometric objects.**

> **_Option 1:_** Create an orthographic drawing of a cylinder, showing front, side, and top views.
>
> **_Option 2:_** Create an orthographic drawing of a symmetric structure made of 13 linking cubes. Show front, side, and top views.

Whether students choose to create the orthographic view of the cylinder or their own structure, they are considering the same big idea about alternate representations of 3-D objects. Some students might shy away from **_Option 2_**, where they have to create their own structure, but others might find this more appealing because it allows them to avoid the issue of the roundness of the cylinder.

Questions applicable to both tasks include:

- *Your shape has symmetry. Do your orthographic views show this? What aspect of them?*
- *Which of the views was easiest for you to create? Why do you think that was?*
- *Which of the views was most difficult for you to create? Why do you think that was?*
- *Do you think you would need all three views to be sure of what the shape looked like?*

✳ **BIG IDEA.** **Locations of objects can be described in a variety of ways.**

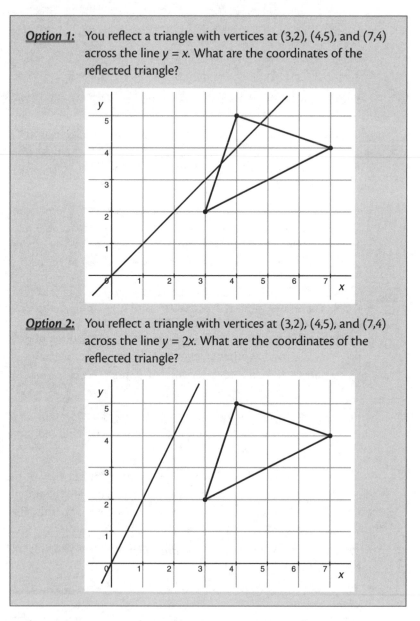

**Option 1:** You reflect a triangle with vertices at (3,2), (4,5), and (7,4) across the line $y = x$. What are the coordinates of the reflected triangle?

**Option 2:** You reflect a triangle with vertices at (3,2), (4,5), and (7,4) across the line $y = 2x$. What are the coordinates of the reflected triangle?

Many students find reflections in the line $y = x$ easier because it simply involves a switch of the two coordinates, for example, (4,5) moves to (5,4). _**Option 1**_ presents a bit of a challenge, though, in that the line goes through the triangle.

Regardless of the option selected, follow-up questions such as these could be asked:

- *Could you have sketched the approximate position of the triangle's **image**? How might that help you when you calculated the coordinate values?*
- *In which quadrant are the new coordinates? Could you have predicted that before you calculated their values?*
- *Did you need to take measurements to decide where the new coordinates would be? If so, which measurements did you need? If not, why not?*

TEACHING TIP. Sometimes it makes sense to add a simple challenge to the "easier" option to ensure that students are always moving forward.

PARALLEL TASKS FOR GRADES 9–12

✹ **BIG IDEA.** Both quantitative and qualitative attributes of a geometric object can impact measurements associated with that object.

> *Option 1:* A triangle has one side of length 10", another of length 4", and another of length 12". What are its angles?
>
> *Option 2:* A triangle has one side of length 12" between angles of 40° and 60°. What are the other angle measure and side lengths?

Although students learn that **SSS** (given three sides) and **ASA** (given two angles and the side between them) are both conditions for congruence (or uniqueness) of triangles, it is often much easier for students to actually construct the ASA triangle than the SSS triangle.

Relevant questions for both tasks include:

- *Which part of the triangle did you draw first? Why did you make that choice?*
- *Could you have made a different choice?*
- *How did you create the triangle?*
- *Could you have created a different triangle with those measurements? Why or why not?*

Variations. Students could be given a choice between an option that provides **AAS** information and another that provides ASA information. Alternatively, students could be given **SSA** information in one case to see that more than one triangle could have been drawn.

Option 1:

D is the **midpoint** of side BC. E is the point where all of the **medians** of the triangle meet. AE is 12" long.

How long is DE? How do you know?

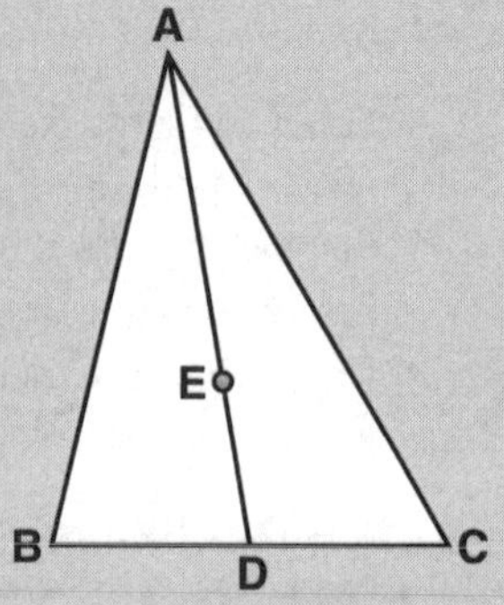

Option 2:

D and E are midpoints of AB and AC, respectively. BC is 12" long.

How long is DE? How do you know?

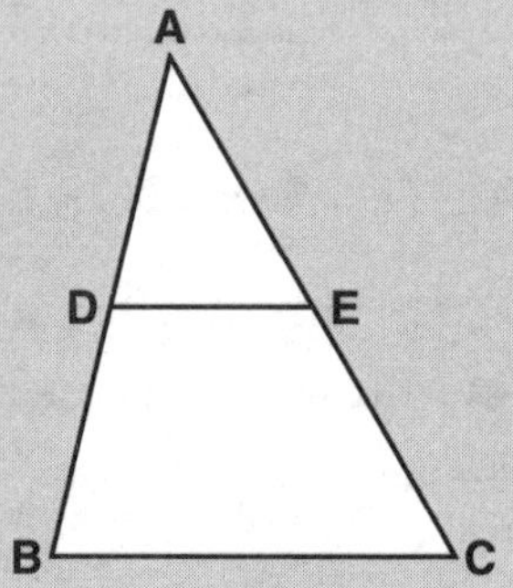

Option 1 focuses on the 2:1 ratio of the parts of a median in a triangle. _Option 2_ focuses on similarity ratios within a triangle. But in both cases, students are using the idea that knowing some measurement information about a triangle yields other information as well.

No matter which option students choose, follow-up questions such as these would be appropriate:

- _Do you think that DE is more or less than 12"? Why?_
- _Do you think that DE is more or less than 3"? Why?_
- _What relationship can you use to help you figure out the length of DE?_
- _How do you know that relationship applies here?_

Option 1:   The points that make up a particular shape are equally distant from the point (1,0) and the line $x = -1$. What does the shape look like? How do you know?

Option 2:   The points that make up a particular shape are $\frac{2}{3}$ as far from the point (1,0) as the line $x = -1$. What does the shape look like? How do you know?

Some students might know that a parabola is defined as the shape made up of points equally distant from a point and a line and choose _Option 1_. Some students might just use sketches. However, both tasks are likely to involve many students in using **distance formulas** on a coordinate grid to determine the appropriate **locus**. Questions applicable to both tasks include:

- *How can you find the points a certain distance from a given point?*
- *How do you measure the distance to the line x = −1? Why does that method work?*
- *Why might you have predicted that your shape would not be a straight line?*
- *How would you describe your shape?*

Variations. Students might be provided different conditions for the required locus of points to meet. For example, the equations of the lines or the given points might change, or the relationship of the distances to those lines or points might differ.

> **Option 1:** Construct △PQR so that PQ = $\frac{3}{4}$QR and ∠PQR = 60°.
>
> **Option 2:** Construct △PQR so that PQ = QR and ∠PQR = 60°.

Option 2 may be more accessible to some students. They can create an isosceles triangle, of any size, as long as ∠PQR is 60°.

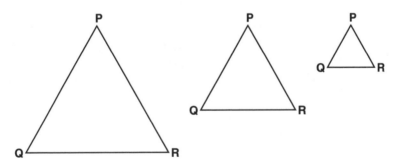

Rarely are students asked to create lengths that are not either equal to or half of another, so working with a new fraction, as required in **Option 1**, might prove to be an interesting challenge for students.

Regardless of the option selected, follow-up questions such as these could be asked:

- *How did you make sure that PQ's length and QR's length satisfied the required relationship?*
- *How did you make sure ∠PQR was 60°? Would there have been ways to do that other than the way you chose?*
- *If someone had drawn a triangle that was different from yours, could it still be correct? What would be different and what would not be different?*

Variations. The relationship between PQ and QR could be changed, as could the measure of ∠PQR. Alternatively, information about a different angle in the triangle could be provided.

> **Option 1:** The vectors [2,x] and [y,5] are perpendicular. What values could x and y have?
>
> **Option 2:** The vectors [2,x] and [y,5] form a 60° angle. What values could x and y have?

There will be some students who draw sketches and hope that their sketches help. Other students might recall that the **dot product** of two vectors must be 0 when the vectors are perpendicular and will select **Option 1**. Still other students might appreciate the challenge of figuring out how knowing the 60° angle plays into the values for x and y; in fact, the dot product of the vectors divided by their magnitudes must be the cosine of the angle, or $\frac{1}{2}$.

Relevant questions for both tasks include:

- *How is information about the angle between the vectors used?*
- *What vector operation did you use to help you solve the problem? Why?*
- *How do you know that there are many possible values for x and y?*

✹ BIG IDEA. **How a shape can be composed and decomposed, or its relationship to other shapes, provides insight into the properties of the shape.**

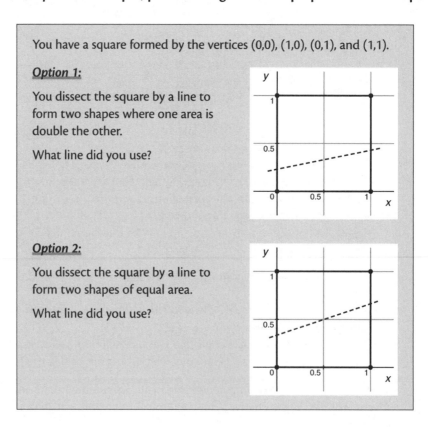

You have a square formed by the vertices (0,0), (1,0), (0,1), and (1,1).

Option 1:

You dissect the square by a line to form two shapes where one area is double the other.

What line did you use?

Option 2:

You dissect the square by a line to form two shapes of equal area.

What line did you use?

Because students could use simple symmetry ideas to help them solve **_Option 2_**, it might be the easier option for some students. Other students, though, might realize that one could solve **_Option 1_** by just drawing a horizontal line at $y = \frac{1}{3}$ and get a very simple solution involving rectangles. They could then be asked whether there are other possible solutions (there are; e.g., $y = \frac{1}{3}x + \frac{1}{6}$).

No matter which option students choose, follow-up questions such as these would be appropriate:

* *What area formulas did you use that helped you?*
* *Why does it make sense that there are many possible solutions?*
* *How could a quick sketch help you see if your line makes sense?*
* *What line did you end up using?*

Option 1: What is the smallest number of triangles into which an *n*-sided regular polygon can be divided? How do you know? What information about the polygon does that provide?

Option 2: Into how many congruent triangles can an *n*-sided regular polygon be divided? How do you know? What information about the polygon does that provide?

Students pursuing **_Option 1_** should come to the realization that joining a vertex to each other possible vertex (to which it is not already joined) provides the minimum number of triangles into which a polygon can be divided. They should be led to see that this would help them determine the total number of diagonals for the polygon.

Those pursuing **_Option 2_** might see the value in locating the center of the polygon and connecting that center to each possible vertex. In this case, students must also recognize why the triangles created are congruent. They should be led to see that this kind of dissection would help them determine the angle size at each vertex.

Questions applicable to both tasks include:

* *Why did every vertex on your polygon need to be connected to another point to solve your problem?*
* *How many lines did you draw? Why that many?*
* *How many triangles did you create? Could you have predicted that number?*
* *Having divided up the polygon the way you did, what new information about the polygon could you now figure out?*

> **_Option 1:_** Divide a regular hexagon into eight identical shapes.
>
> **_Option 2:_** Divide a regular hexagon into four identical trapezoids.

Each of these options requires the student to consider properties of shapes and to use those properties to decompose the regular hexagon.

Option 2 might be easier for the student who thinks of using non-isosceles trapezoids. **_Option 1_** might be easier for the student who thinks of first dividing the shape into six equilateral triangles, then subdividing each of them into four identical triangles, and finally grouping sets of three triangles to create trapezoids.

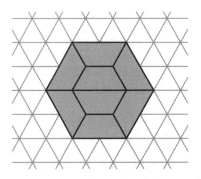 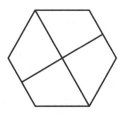

In either case, students might use dynamic geometry software to help them solve the problem.

No matter which option students choose, follow-up questions such as these would be appropriate:

* *How do you know the shapes will have straight sides?*
* *Does the fact that the shapes have to be identical affect how you start the problem?*
* *What are some easy ways to divide up a regular hexagon?*
* *Does your answer depend on the fact that the hexagon was regular? How?*
* *What strategy did you use to come up with your dissection of the hexagon?*

> **_Option 1:_** Is it always possible to divide an obtuse triangle into acute triangles? Explain your thinking.
>
> **_Option 2:_** Is it always possible to divide an isosceles triangle into other isosceles triangles? Explain your thinking.

In this situation, **_Option 2_** is considerably simpler than **_Option 1_**. With **_Option 2_**, a student merely needs to use the concept of similarity; the student can draw 4 similar triangles within the original triangle. This is always possible by drawing a line parallel to the base at the midpoints of the two equal sides and then dividing the bottom trapezoid into three congruent triangles.

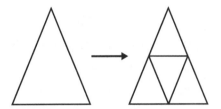

Option 1 requires much more sophisticated thinking. A student needs to know that if a line is drawn to cut the obtuse triangle, it cannot go all the way to the other side or another obtuse triangle or a right triangle, but not an acute triangle, would be formed.

An obtuse triangle can always be divided into seven or more acute triangles (five triangles forming a pentagon inside the triangle, to ensure that all of the angles that meet at the center of the pentagon are acute, and the two other triangles at the ends).

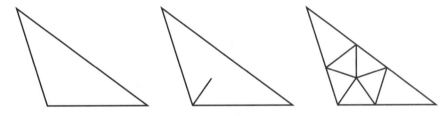

Regardless of the option selected, follow-up questions such as these could be asked:

- *Do you think you could solve the problem using just two triangles? Why or why not?*
- *How many measurements for each triangle will you need to take to see if your triangles are correct?*
- *Could all of the triangles inside be similar to the original triangle? Why or why not?*
- *What strategies did you use?*
- *What information about triangles did you use to help you apply your strategy?*

✹ **BIG IDEA.** Locations of objects can be described in a variety of ways.

> *Option 1:* A plane goes through the points (5,1,5), (6,5,3), and (10,2,14). What is its equation?
>
> *Option 2:* A plane goes through the points (6,9,8), (4,17,8), and (12,15,8). What is its equation?

In both situations, three points are provided and a unique plane is determined. Students who are observant might realize that the fact that all three z-coordinates are 8 in *Option 2* guarantees that the plane is $z = 8$. In either case, though, students could solve a **system of linear equations**—three equations in four unknowns—to solve for the equation of the plane. For example, for **_Option 1_**, if the plane is of the form $ax + by + cz = d$, the system of equations is:

$$5a + b + 5c = d$$
$$6a + 5b + 3c = d$$
$$10a + 2b + 14c = d$$

Regardless of the option selected, follow-up questions such as these could be asked:

- *Could there be more than one plane that goes through the three points listed? How do you know?*
- *What does the equation for a plane look like?*
- *How would you know if the given points were on the plane you selected?*
- *How did you determine the equation of the plane?*

SUMMING UP

MY OWN QUESTIONS AND TASKS

Lesson Goal: _____   Grade Level: _____

Standard(s) Addressed:

Underlying Big Idea(s):

Open Question(s):

Parallel Tasks:
Option 1:

Option 2:

Principles to Keep in Mind:
- All open questions must allow for correct responses at a variety of levels.
- Parallel tasks need to be created with variations that allow struggling students to be successful and proficient students to be challenged.
- Questions and tasks should be constructed in such a way that will allow all students to participate together in follow-up discussions.

The five big ideas that underpin work in Geometry were explored in this chapter through 46 examples of open questions and parallel tasks, as well as variations of them. The instructional examples provided were designed to support differentiated instruction for students at different developmental levels, targeting two separate grade bands: grades 6–8 and grades 9–12.

Sometimes a student struggling with Number and Operations or Algebra performs well in Geometry or vice versa. Rather than assuming that certain students need scaffolding and others do not based on their performance in Number and Operations or Algebra, teachers can use differentiated questions and tasks, as suggested, to allow students to work at an appropriate level in Geometry.

The examples presented in this chapter are just a beginning. Other questions and tasks might be created by, for example, analyzing and performing a variety of transformations, especially on algebraic graphs or by examining additional concepts in vector geometry. A form such as the one shown here can serve as a convenient template for creating your own open questions and parallel tasks. The Appendix includes a full-size blank form and tips for using it to design customized teaching materials.

Measurement

DIFFERENTIATED LEARNING ACTIVITIES in measurement are derived from applying the NCTM process standards of problem solving, reasoning and proof, communicating, connecting, and representing to content goals of the NCTM Measurement Standard, including

- understanding measurable attributes of objects and the **units**, systems, and processes of measurement
- applying appropriate techniques, tools, and formulas to determine measurements (NCTM, 2000)

TOPICS

Before beginning the task of differentiating student learning in measurement, it is useful for teachers to have a good sense of how the topics in the strand develop over the grade bands. The NCTM *Curriculum Focal Points* (NCTM, 2006), which suggest what mathematical content should be the focus at each grade level through grade 8, were used as the basis for recommendations made in this resource for grades 6–8; the NCTM *Principles and Standards for School Mathematics* (NCTM, 2000) helped form the basis for the material for the higher grades. For a teacher at a particular grade level, it can be helpful to be aware of where students' learning is situated in relation to what learning has preceded the present grade band and what will follow.

Grades 6–8

Within this grade band, students continue to solve problems involving length, area, and volume. They develop and use formulas for surface areas and volumes of prisms and cylinders and for areas and circumferences of circles.

Students in this grade band begin to recognize the relationships between measurements of similar shapes, and they use the Pythagorean theorem to simplify the measurement of lengths of the sides of right triangles.

Grades 9–12

Within this grade band, students more explicitly consider issues of **precision, accuracy**, and **measurement error**; apply formulas for surface area and volume to a broader range of 3-D shapes; use concepts of limit to estimate measures; and use **unit analysis** to make sense of **derived measurement** computations.

They continue to consider the choice of appropriate benchmarks, and they use area formulas and notions of similarity to analyze measurements in 2-D situations, including the study of trigonometry.

THE BIG IDEAS FOR MEASUREMENT

Coherent curricula in measurement that meet NCTM content and process standards (NCTM, 2000) and support differentiated instruction can be structured around the following big ideas:

- A measurement is an explicit or implicit comparison.
- The unit or tool chosen for a measurement can affect its numerical value as well as the precision of the measure.
- The same object can be described using different measurements; sometimes the measurements are related, and other times they are independent.
- Knowing the measurements of one shape can sometimes provide information about the measurements of another shape.
- Measurement formulas allow us to rely on measurements that are simpler to access to calculate measurements that are more complex to access.

The tasks set out and the questions asked while teaching measurement should be developed to evoke these ideas. The following sections present numerous examples of application of open questions and parallel tasks in development of differentiated instruction in these big ideas across two grade bands.

OPEN QUESTIONS FOR GRADES 6–8

OPEN QUESTIONS are broad-based questions that invite meaningful responses from students at many developmental levels.

✴ **BIG IDEA.** A measurement is an explicit or implicit comparison.

> Describe a situation in which you might want to estimate either the circumference or the area of a circle but would not want to calculate it exactly.

It is important that students recognize that often an estimate provides sufficient information to solve a problem. Many students are uncomfortable with estimation and choose to rely on more exact calculations, estimating after the fact. Questions like the one provided force students to think about the fact that exact calculations are not always required.

To answer the question, a student might suggest that an estimate of an area is enough to decide how much paint to purchase to paint a circular target, because the paint can be thinned a bit if needed or because cans of paint only come in certain sizes anyway.

If students struggle to come up with a response, the teacher could provide scaffolding with questions such as these:

- *What circular objects might you be finding the circumference or area of?*
- *For what purpose would you need these measurements?*
- *Can you think of an example where you would only estimate a length?*
- *Could a similar situation apply to area?*

> A hexagonal prism is a lot bigger than a cube with the same height. What might the bases of the two objects look like?

This question is very open in that the student can decide what type of hexagon to use, can use personal judgment to interpret the phrase "a lot bigger," and can describe the bases either using numbers, pictures, or words. A student might, for example, simply draw any large hexagon and a much smaller square. Or a student might, much more precisely, determine the areas of a regular hexagon and square and compare them. Either student is showing an understanding of the notion that the volume depends on both height and base area.

Variations. The question could be varied quite simply by changing the shapes to be compared from hexagonal prisms and cubes to other pairs of shapes.

> You say that a car is driving at a rate of 55 mph. What might you think about to have a sense of how fast this is?

Students need to recognize that a measurement (such as miles per hour) is meaningful only if the user can compare it with something he or she already knows. This question is open because the students have many choices about how to make this comparison.

For students having difficulty getting started, questions such as these could be helpful:

- *In what situation might this speed be likely?*
- *Is this faster than a bicycle speed? A lot faster?*
- *How far could you go in an hour at this speed as compared with another speed of your choice?*

> What strategies can you use to tell whether one pair of cities is closer together than another pair? Why does each strategy work?

To compare measurements of the distances between pairs of cities, students might choose to use numerical values (e.g., distances looked up on the Internet or in a chart in a road atlas), comparing those values as numbers. They might also use a map and compare ruler-measured distances between the cities, or they might even cut strings to the sizes of the spaces between the pairs of cities on the map and directly compare the string lengths. The latter method is possible for virtually any student.

No matter what strategies are used, it is important for the teacher to bring out that in each case, two measures are being compared.

✸ BIG IDEA. **The unit or tool chosen for a measurement can affect its numerical value as well as the precision of the measure.**

> You measure the area of a room as 90 yd². Suppose you measured in square feet instead. What do you know about how many square feet it would be?

Some students will know that you can multiply the 90 by 9 (not 3) to determine the new measurement. However, this question is open in that a student might simply say that the new number will be greater because a smaller unit is being used. Another student might say that the area is the same because the room did not change size; that student, too, will be correct.

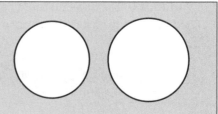

> You want to compare the circumferences of these two circles.
>
> What unit and what tool would you use? Why?

Initially, some students are likely to focus on the fact that it is hard to measure round objects with rulers, and they might suggest using a string and then placing that string against a ruler. Other students will realize that they can use a ruler or **calipers** directly, because by measuring the diameter they can easily calculate the circumference by using a formula.

Because the circles are close in size, students should realize they need a tool with units that are fairly small, such as a ruler that has fractions of inches clearly marked. Some students will decide that any unit can be used as long as the tool they are using allows them to use a sufficient number of decimal places. For example, the circumferences could be compared in terms of thousandths of a meter.

Variations. Variations of the question could be created by asking about the areas of the circles instead of their circumferences.

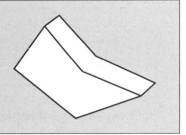

You want to determine the volume of this pentagonal prism.

What might you do to get a good estimate?

Using an irregular shape opens up this question. If, for example, a rectangular prism had been shown, most students would use a known formula. Because this unusual shape was used, students are much more likely to think about options, realizing that trying to build the shape with cubes and count the cubes would lead to too imprecise a value. In addition, no specific measurements are reported, which leads students to think more about process rather than launch right into calculations.

Some students will realize that one strategy might be to use water displacement, that is, to immerse the shape in a liquid and determine the displaced volume by relating it to the rise in the liquid measure. Other students will think of various ways to determine the area of the base and the height of the prism and use this alternative.

TEACHING TIP. Sometimes presenting a less familiar situation does not actually make a question more difficult but, rather, makes it more open.

✹ **BIG IDEA.** The same object can be described using different measurements; sometimes the measurements are related, and other times they are independent.

The length of one side of a right triangle is 10 cm. What might the lengths of the other two sides be?

Some students will use the Pythagorean theorem, $a^2 + b^2 = c^2$, substituting the value of 10 for either a, b, or c and calculating the other values. Other students will simply draw a right triangle, marking either one **leg** or the **hypotenuse** as 10, and reading off the measurements of the other sides of the triangle.

The question is open in that it allows for both of these approaches. Even the student who simply measures without using the formula will benefit from the discussion of how simple the problem becomes when the formula is used.

Variations. Instead of providing information about one side length of a right triangle, students might be provided with one side length of an isosceles triangle. This is a more complex problem, suitable for mathematically stronger students.

> The circumference of a circle is more than 10". What else do you know about the circle?

Students might approach this open question in different ways. Some students might simply state that the circumference is more than 9". Some might focus strictly on geometric properties that have nothing to do with size, for example, that it is round. Other students will focus on what other measurements of the circle might be, for example, that the radius is a little more than $1\frac{1}{2}$" or the diameter is more than 3". By allowing these alternative interpretations of the question, virtually every student can be successful.

If students do not know how to begin, the teacher could ask scaffolding questions such as:

- *What geometric attributes would the circle have?*
- *Do those depend on the size of its circumference?*
- *What attributes might depend on that size?*

Variations. The question could easily be adjusted by changing the measurement, changing the information about the circle given (e.g., giving the diameter or area instead), or by changing "more than" to "less than."

> One of the measurements of a circle is 6 units. Draw and describe or show at least one other measurement of the circle.

Because students have freedom to choose which measurement is 6 units, this question becomes accessible to many students. A student might, for example, choose the radius to be 6" and calculate the diameter, or vice versa. Or a student might choose the radius or diameter to be 6" and calculate the circumference or area. Some students might choose the 6 units to be the circumference and calculate the diameter, radius, or area. Because the term *units* was used in the question, rather than a specific linear unit such as inches, students might also interpret the units to be area units and determine the radius, diameter, or circumference.

TEACHING TIP. When there are both more complex and less complex ways to interpret and respond to a question, a teacher can intervene when he or she thinks it is in the best interests of a particular student to seek a more or less complex response.

> Two points on a circle are joined. The line segment joining them is 5"
> and does not go through the center of the circle. What do you know
> about other measurements of the circle?

Most students will draw a picture to approach the
problem. Some students will observe that the diameter
of the circle must be greater than 5", so the radius must
be greater than $2\frac{1}{2}$". Other students might note that the
circumference must be more than 15" and that the area
must be more than 19 in^2. Some students might simply
state that the circle cannot be too tiny or the 5" segment
would not be possible.

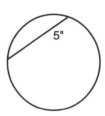

> You know that a cylinder has a height of 10" and a base
> circumference of 7". Which other measurements of the cylinder can
> you now calculate with certainty? Are there any measurements of the
> cylinder you cannot be sure of?

This open question does not ask about a specific measurement but leaves the
student free to think about which measurements they might want to consider.
Some students will think about all of the measurements that can be determined
and others will think of only some of the measurements.

For students who struggle to come up with a response, questions that could
be asked include:

- *Do you know the distance around the middle of the cylinder? Why or why not?*
- *What could you measure on the base of the cylinder?*
- *Which of the given measurements would help you figure that out? How?*

✳ BIG IDEA. **Knowing the measurements of one shape can sometimes provide
information about the measurements of another shape.**

> The string used to make a circle is just a little
> longer than the string used to make the
> square shown.
>
> How wide is the circle?

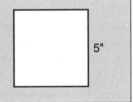

This question, like many other open questions, is deliberately vague. The term
a little longer can be interpreted by students in different ways.

Students might realize that *a little longer* in a situation where the square is only 5" wide could be significantly different from *a little longer* in the context of a 5' square. The term *wide* is also somewhat vague—a student might interpret this to mean the diameter, might choose to identify the radius, or might simply describe an object that is of the appropriate size without referring to its measurements.

The question allows access to either a student who physically creates the shape or one who chooses to use formulas.

If students have difficulty getting started, the teacher could provide scaffolding with questions such as these:

- *Is the string 5" long?*
- *What would you have to do to figure out how long the string is?*
- *What measurement of the circle would the string's length tell you the most about? Why?*

TEACHING TIP. Using softer language—such as *a little longer, a little shorter,* and so forth—offers the opportunity for success for many who typically struggle when more exact answers are required.

A triangular prism has a volume that is just slightly less than the volume of this rectangular prism.

What could be the dimensions of the triangular prism?

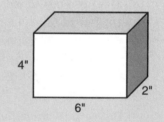

By using simple values for the measurements, the question becomes accessible to a broader range of students. Many students will realize that the volume of this shape is 48 in³. What they are likely to realize in the case of the triangular prism is that the critical measurements for determining the volume are not the edge lengths, but the area of the base (depending on the triangle's base and height) and the height of the prism. Because the nature of the triangular prism is left open, a student could choose a right triangular base, which might make the visualization of the problem simpler.

For example, a student might decide to use the same height of 4", but, in order to get a base area of a bit less than 12 in², might choose a triangular prism with a side length of 11" and an altitude of 2". Other students might more broadly interpret the term *a little* and use an even smaller triangular base that they could determine through experimentation. With either approach, students would be using the notion that the base area and the height of a prism are relevant in determining its volume.

> A right triangle has one 50° angle and one side length of 10 cm. A similar triangle has one side length of 16 cm. What might all of the triangle side lengths in both triangles be?

This question allows students to flexibly work with numbers. The student could select either of the legs or the hypotenuse of the original triangle to be 10 cm. The student could then calculate the other side lengths and angles, matching the 16 cm side in the similar triangle to either the 10 cm side or one of the other sides.

For example, the solution could involve multiplying the side lengths 10 cm, 8.38 cm, and 13.05 cm in the first triangle by 1.6 so that the vertical side is 16.

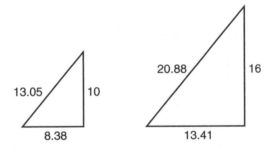

An alternative is to multiply each side length by 1.23 so that the hypotenuse is 16 or to multiply each dimension by 1.91 so that the horizontal side is 16. Additional solutions can be created if the measurement of 10 is applied to a different side length in the first place.

✳ **BIG IDEA.** Measurement formulas allow us to rely on measurements that are simpler to access to calculate measurements that are more complex to access.

> What do you know about the number π?

Students meet the number π as they learn about the formulas for the measurements of a circle. Some students just think about pi as a symbol you use in those formulas, but others have a broader understanding of its value and what it represents.

There are many possible answers a student might give:

It is more than 3, but less than 4.
It is the ratio of the circumference to diameter of any circle.
It is the ratio of area to the square of the radius of any circle.
It is less than 5.
It is more than 1.
It is a number, but it's not a fraction.
It is a letter in the Greek alphabet.

> The label on a cylinder-shaped can is cut vertically from the top of the can to the bottom. The label is then stretched out. The area of the label is more than 200 in².
>
> What do you know about the width of the can?

Some students may need to make a drawing or use a physical model to realize that the label shape is a rectangle with length the circumference of the base of the can and width the height of the can.

OPEN QUESTIONS FOR GRADES 9–12

✳ **BIG IDEA.** A measurement is an explicit or implicit comparison.

> You know that the volume of a particular square-based pyramid is 12 ft³. Choose another 3-D shape to compare its size to, and then describe the comparison.

This question is very open in that the student is allowed to choose any alternate 3-D shape with which he or she is comfortable. In fact, a student might choose a cube that is 1 ft on a side and say that the pyramid is 12 times as large as that cube. Another likely response might be comparing the pyramid to a prism with the same base and height, which would have a volume of 36 ft³. There are many other possible comparisons as well.

Variations. Rather than a pyramid, a cone could be used. In that way, the student might compare to the related cylinder, to a benchmark cube, or to some other shape.

> How are sin θ and cos θ alike? How are they different?

By asking this open-ended question, the teacher encourages the student to consider as many things as they know about these two ratios. In terms of similarities, students might consider the values they can take on (i.e., both go from 0 to 1 for right triangles), the fact that they both describe ratios of side lengths of right triangles, or the fact that the denominator for each ratio is the hypotenuse of the right triangle.

Differences that might be mentioned include the fact that different side lengths of a right triangle are used to determine them, the fact that one increases and the other decreases as angles increase from 0° to 90°, or perhaps the values they take on for a specific angle.

Variations. The question could be varied by using other combinations of **trigono-metric ratios**, for example, sine and **tangent**, tangent and **cotangent**, and so forth.

✸ BIG IDEA. **The unit or tool chosen for a measurement can affect its numerical value as well as the precision of the measure.**

> Describe how far around you go if you start at point P and go around the circle counter-clockwise until you get back to point P.
>
> Is there more than one way to describe that distance?

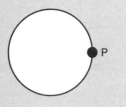

In the higher grades, students are introduced to **radians** as an alternate way to describe circular motion. In fact, a radian is a real number that is the ratio of the distance around a circle to the radius of a circle, although many students think of it as an angle equivalent.

By asking this open question, students can think of the motion as a distance around the circle (its circumference), as a 360° turn, or as 2π radians. By allowing for all of these possibilities, a richer discussion ensues. Students can think about why the measurement numbers are different and how they relate to one another.

> How can you closely estimate the shaded area?

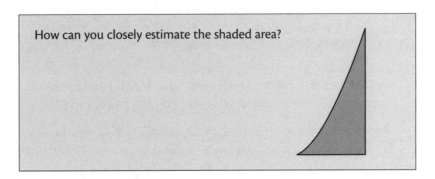

To get ready for calculus, students start to think about limits. In this case, students will likely think of estimating the area by using known smaller shapes. Those who use the sum of narrower and narrower shapes for the estimates will have a closer estimate of the area than other students. By allowing students the choice of how to estimate the area closely, it is likely that a class discussion will bring out the idea of limits.

✳ **BIG IDEA.** **The same object can be described using different measurements; sometimes the measurements are related, and other times they are independent.**

> Sam argues that there is no need for knowing trigonometry—if you want the side lengths of a triangle, just measure them. Do you agree or disagree? Explain.

Asking students to agree or disagree with a statement and explain their reasons can encourage them to think about broader mathematical issues. In this situation, a student could legitimately argue that there is no need for trigonometry in situations where the triangles are small and directly measurable. However, other students will, no doubt, bring up situations where measurement, for one reason or another, might be difficult to accomplish. This helps establish, for all students, the reason they are studying this topic.

For students who do not know how to begin, questions such as these could be helpful:

- *Would you have to consider the tools you have available?*
- *Would your answer be different depending on the angles in the triangle?*
- *Would it be different depending on the side lengths of the triangle?*

> The cosine of an angle in a right triangle, rounded to the nearest thousandth, is 0.707. What might be the dimensions of the triangle?

This question is open in that a student can think of any one of a set of similar triangles. In fact, an isosceles right triangle of any size works.

Variations. Instead, a situation could be presented in which a range of answers is possible, beyond using similar shapes. For example, the student might be asked to create a triangle in which one of the angles has a cosine of less than 0.1.

TEACHING TIP. Starting with the answer allows a teacher to create many questions. In this case, rather than drawing a triangle and asking for a trigonometric ratio, the ratio is given and the triangle is created.

> A construction company wishes to advertise the new and innovative wheelchair ramp it has created. What is one way they could describe the steepness of the ramp?

By asking the question in this way, students have a choice about considering angles, trigonometric ratios, side proportions, or slopes to describe the ramp. Asking this type of open question allows for a rich discussion of how all of these descriptions are ultimately different representations of the same idea.

If students struggle to come up with a response, the teacher could ask scaffolding questions such as:

- *Do you think the ramp might be as steep as the one shown at the right? Explain your thinking.* 
- *How might you use words to compare that steepness to a different one?*
- *How might you use numbers to compare that steepness to a different one?*

> The sine of an angle in a right triangle is fairly large. What might the two acute angles of the triangle be? Explain.

Rather than giving the value of the sine of the angle and having students use calculators or computers to determine the angle size and the values of other angles in the triangle, this more open question forces students to think about what it means to have a large sine.

The word *large* is fairly vague, leaving the question open. It allows a student who is not yet comfortable with using the **inverse sine** function to still respond meaningfully, using conceptual notions of what the sine is and does. The question also allows for a discussion of the fact that because the sines of angles cannot be more than 1, a large sine is probably more like 0.8 or 0.9 than it is a number such as 1,000, which, in absolute terms, might have been the student's first thought about what a large number is.

Class discussion could also lead students to recognize that if one angle in a right triangle has a large sine, the other acute angle must have a small sine, closer to 0.

> Aaron said: The sine of acute angle A of a right triangle cannot be greater than the cosine of angle B in the same triangle. Do you agree? Explain.

This sort of question lends itself to experimentation. A student might think about different right triangles and different possible combinations to respond to the question. Alternatively, a student might reason that the sine of angle A must be equal to the cosine of angle B because they are **complementary angles**. For this reason, Aaron's statement is true.

For students having difficulty getting started, questions that could be asked include:

- *What does an acute angle look like?*
- *What else could it look like?*
- *What could it not look like?*
- *Could angle B be* **obtuse***?*
- *How do you calculate the sine of each? The cosine of each?*

> When and why is it useful to know what the sine of an angle is?

Often students remain somewhat oblivious to the usefulness of the mathematical topics they are taught. An open question such as this one is likely to generate a class discussion about the usefulness of trigonometry as an area of study.

Some students might suggest contextual situations in which knowing the sine of an angle could be useful. For example, they might suggest a situation in which one would want to know the height of an object and using the sine might help if the hypotenuse length is known.

Other students might suggest mathematical situations. For example, a student could suggest that it would be useful to know the value of the sine of an angle if a person wanted to determine its cosine.

Still other students might suggest very different ideas, such as understanding what is happening when certain buttons on a scientific calculator are pressed or for passing a test in trigonometry.

All of these students are right. Given that the purpose of open questions is to generate mathematical discussions to which all students can contribute, this question might be somewhere near the end of a trigonometry unit.

Variations. A variation of the question could be created by asking about a different trigonometric ratio—cosine or tangent, for example.

> The triangle below represents a fairly large backyard area. Choose lengths for the missing sides and a measurement for the missing angle. Then choose a price for one square yard of sod to cover the area. Decide how much it would cost to plant this triangular patch of land with sod.
>
>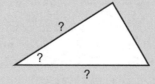

Students are likely to estimate the size of the missing angle in the diagram and the relative lengths of the two unknown sides. They are also likely to do some investigation to find out what reasonable sod prices might be and to estimate reasonable yard dimensions. All of these are useful real-life skills.

Some students might make the problem slightly easier by choosing "friendly" numbers for side lengths, such as 20 yd or 30 yd, and prices such as $1 per square yard, as opposed to numbers more challenging to use in calculations.

TEACHING TIP. One simple strategy for creating open questions is to allow the student to choose some of the values in the question. This approach allows the student to focus on the underlying concept rather than becoming unnecessarily bogged down in calculations. It also allows for a more interesting conversation after the fact, because the answers of different students might reasonably be compared.

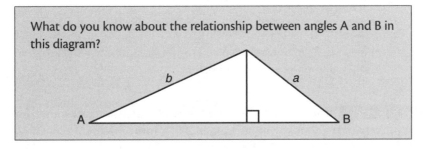

What do you know about the relationship between angles A and B in this diagram?

Some students might simply observe that if angle B > angle A, then the opposite side b > the opposite side a; this can be seen visually. Other students might realize that the height of the triangle can be expressed as either $b \sin A$ or $a \sin B$, Therefore, this open question can be used to help students derive the **law of sines**, that is, $a/\sin A = b/\sin B$.

The cosine of an angle is a lot more than its sine. What do you know about the angle?

This question is deliberately vague in its use of the phrase *a lot more*. With the question stated in this way, many more students can be successful; they have control over what they call *a lot*. Through the course of the discussion, however, it is likely that the students will see that angles closer to 0° than to 90° have cosines greater than their sines.

Variations. Clearly, one variation of this question could be to change the phrase *a lot more* to the phrase *a lot less*. Other variations could involve using other trigonometric ratios. For example, students could be told that the sine is much less than the tangent and asked what they know about the angle.

> It is easy to calculate the sine of a certain angle even without a calculator. What might the angle be?

An opinion question such as this one has a great deal of potential for including all students in the class. Responses could range from using 0° and 90°, extreme values, to using other familiar angles such as 45° or 60°, to a suggestion that even though a calculator is not used, maybe a table could be. In that case, any angle less than 90° might work.

If students do not know how to begin, the teacher could provide scaffolding with questions such as these:

- *Which angle measures are easiest for you to visualize?*
- *How does your mental picture of the angle help you see what its sine might be?*
- *Does it depend on how big you make your triangle whether you can visualize the value of the sine?*

Variations. It could be interesting to vary the question by insisting that the angle be greater than 90° once students have met situations in which trigonometric ratios are applied to obtuse or **reflex angles**.

✳ BIG IDEA. **Knowing the measurements of one shape can sometimes provide information about the measurements of another shape.**

> How could you use trigonometry to help you figure out the height of a tall building?

This open question takes students out of the realm of the specific to describing a more general real-life application involving trigonometry. Students might create a fairly simple situation in which they use an **angle of elevation** and distances to a building or a much more complex situation where they use two angles of elevation from different positions, but only one ground distance.

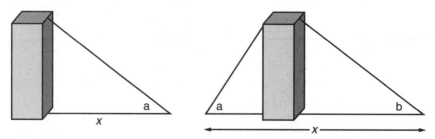

No matter what kind of situation students might create, they are thinking about how trigonometric ratios are used to determine unknown measurements given known measurements.

> Choose equations for two lines that are neither parallel nor perpendicular. Determine the angle between them without using a protractor.

One student might use simple lines that intersect at the origin and have simple slopes. For example, the lines might be $y = x$ and $y = 2x$. (See Graph A below.) The student might realize that the lower line makes a 45° angle with the horizontal. The student might realize that the angle to the horizontal for the upper line has a tangent of 2, so the angle must be close to 64° and the difference, or angle between the lines, must therefore be 19°.

In contrast, another student might first choose an angle and then determine the equation of the line. (Even though the instructions say to choose the lines first, this should be allowed.) The student might draw $y = 0$ and $y = x$. The student knows the slope of the line $y = x$ is 1, and so the angle is 45°.

Yet another student might perform a much more complicated task. For example, the student might draw $y = 5 - 2x$ and $y = x + 4$. (See Graph B below.) This student has a lot more work to do to figure out the best way to calculate the angle.

Graph A	**Graph B**
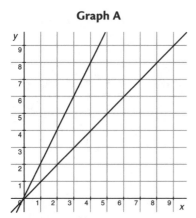	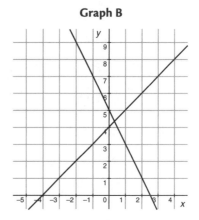

For students who struggle to come up with a response, questions such as these could be helpful:

- *How were you sure your lines were neither parallel nor perpendicular?*
- *Did you draw a picture first? Was that helpful?*
- *Could you estimate the angle before you calculated it? Why might that be useful?*

TEACHING TIP. Sometimes students ignore part of the instructions or misunderstand something. Rather than demanding that they do exactly what was wanted, it is much more inclusive to allow the misinterpretation to stand, so long as nothing wrong is actually said. This way, students are much more likely to volunteer again.

Open Questions for Grades 9–12

✳ **BIG IDEA.** **Measurement formulas allow us to rely on measurements that are simpler to access to calculate measurements that are more complex to access.**

> You know the height of a cone and the radius of its base. Would you calculate the surface area of that cone using a formula or using a different method? Explain your choice.

Opinion questions have a valuable place in mathematics instruction. They provide a way to give students voice. In this case, one student might argue that the formula is easier because measuring with a grid would be cumbersome or inexact. Another student might argue that it is so hard to figure out the **slant height** of the cone that it would be easier to just take it apart and estimate the area using a grid.

TEACHING TIP. It is important to ask some questions in math class that allow students to offer an opinion. Voicing an opinion on a mathematical subject often has a significant and positive impact on their view of the kind of subject math is.

> Jason said that the **law of cosines** makes sense because the closer a triangle is to being a right triangle, the closer the sum of the squares of the two short sides comes to matching the square of the long side. Do you agree? Explain.
>
>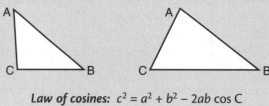
>
> *Law of cosines:* $c^2 = a^2 + b^2 - 2ab \cos C$

This question is open in that the explanations could be visual or algebraic. In either case, students need to think about the notion that if an angle is 90°, its cosine is 0, and thus the law of cosines becomes the Pythagorean theorem. If the angle is close to 90°, the cosine is close to 0, and the term $2ab \cos C$ becomes negligible.

An interesting discussion could ensue as a result of students suggesting that if a and b are pretty small, then the term $2ab \cos C$ is still small even if the angle is not close to 90°.

PARALLEL TASKS FOR GRADES 6–8

> **PARALLEL TASKS** are sets of two or more related tasks that explore the same big idea but are designed to suit the needs of students at different developmental levels. The tasks are similar enough in context that all students can participate fully in a single follow-up discussion.

✸ BIG IDEA. **A measurement is an explicit or implicit comparison.**

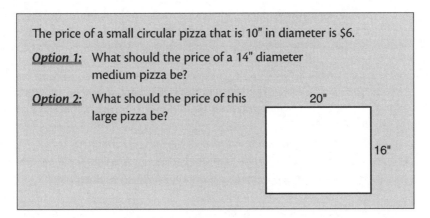

The price of a small circular pizza that is 10" in diameter is $6.

Option 1: What should the price of a 14" diameter medium pizza be?

Option 2: What should the price of this large pizza be?

These tasks provide work in proportional reasoning along with work related to areas of shapes. Although, for either option, students are likely to calculate the area of one circle (the small pizza), in one option they can use a simple rectangle as their second shape, whereas the other option requires a calculation involving another circle. It is possible for students to simply estimate rather than perform any exact calculations.

Relevant questions for both tasks include:

* *Why might you calculate the area of the pizza to answer the question? Do you have to?*
* *How could you estimate the area of the small pizza?*
* *How did you estimate or calculate the area of the bigger pizza?*
* *Could you have predicted whether the price would be more or less than twice $6? Why or why not?*

Variations. Students could be allowed to choose their own price for the small pizza so that it would make sense for one of the other pizzas to cost about $15 or $20.

> *Option 1:* The hands of a clock are 60° apart. What time might it be? List as many possibilities as you can.
>
> *Option 2:* The hands of a clock are 45° apart. What time might it be? List as many possibilities as you can.

Although many students will be attracted to *Option 1*, where one simple response is 2:00 and another is 10:00, these students can be encouraged to look for other answers as well. Students might even consider the possibility of a second hand.

Whichever task is selected, students need to think about the fact that the hour hand moves only 0.5° per minute, but the minute hand moves 6° per minute.

No matter which option students choose, follow-up questions such as these would be appropriate:

* *How do you know that the hands are 90° apart at 3:00 but not at 3:30?*
* *Do the minute hand and hour hand move the same number of degrees in an hour?*
* *Do they move the same number of degrees in a minute?*
* *If the hands are a certain number of degrees apart at one time, are they the same number of degrees apart an hour later? A minute later?*
* *Which times did you choose?*
* *How do you know you're correct?*

✳ **BIG IDEA.** **Knowing the measurements of one shape can sometimes provide information about the measurements of another shape.**

> A circle has a diameter of 8". A new circle is created to meet one of the conditions below. What is the diameter of the new circle?
>
> *Option 1:* The new circumference is 5 times as big.
>
> *Option 2:* The new area is $\frac{1}{4}$ as big.

Students have a choice of working with linear circumference measures or quadratic area measures. The intent is for students to learn that when a linear dimension is multiplied by *n*, so are other linear dimensions, whereas the area is multiplied by n^2. Many students will find the linear comparisons easier to understand than the quadratic comparisons. Even though any particular student is looking at only one type of situation, the class discussion will involve both.

Questions applicable to both tasks include:

* *Is the new diameter likely to be greater than or less than the original?*
* *Is the circumference of the new circle going to be greater than or less than the original one? What about the area?*

- *Did you need to calculate the circumference or area of the new circle to answer the question?*
- *How many circles did you try before you thought you could be sure?*

Variations. Students could be told that an area is doubled or quadrupled and asked what happens to the radius, the circumference, or the diameter.

TEACHING TIP. It is always useful to explore "what if" questions—where one assumption or constraint in a situation is altered to see its effect on other aspects of the situation.

✹ **BIG IDEA.** Measurement formulas allow us to rely on measurements that are simpler to access to calculate measurements that are more complex to access.

> Two circles have different diameters.
>
> _Option 1:_ Can they have the same area?
>
> _Option 2:_ Can they have the same circumference?

The purpose of this question is to ascertain whether students really understand the measurement formulas associated with circles. Whether a circumference or area formula is being used, the objective is to get students to realize that there is only one value of r (or d) associated with a given circumference or area. They also need to realize that they have to apply reason and cannot tell, from just a few examples, whether the possibility they are looking for exists or not.

Some students will find the circumference formula easier to deal with because there is no necessity for calculating a radius and then squaring it.

Students are likely to learn that rectangles with different lengths can still have the same area. Engaging in the kind of thinking required in this question helps them understand why this is not true for regular figures, such as squares, regular hexagons, or circles.

Regardless of the option selected, follow-up questions such as these could be asked:

- *Did you try an example?*
- *How many examples would you have to try to be sure?*
- *What measurement formula did you use to help you?*
- *Why did you use that formula?*
- *How can using the formula help you answer the question?*

> **_Option 1:_** Draw a shape with an area of 2π square units.
>
> **_Option 2:_** Draw a shape with an area of 9π square units.

This question is designed so that students can use any shape they want, although they are likely to use circles. If they do use circles, they can use a radius that is a whole number (for **_Option 2_**) to make the problem more accessible. Students choosing **_Option 1_** need to use a radius of about 1.4, whereas those using **_Option 2_** can choose a whole number radius of 3.

The question focuses on the stated big idea because it is through using formulas that the problem becomes much easier to solve.

Relevant questions for both tasks include:

- *Why you think the area involved π?*
- *What formula for the area of a circle helped you solve the problem?*
- *How did the formula help?*
- *Why wasn't the coefficient of π the radius of the circle?*
- *Did your shape have to be a circle?*

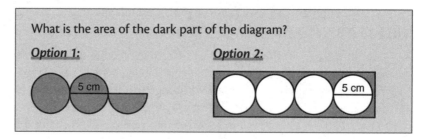

Each option involves calculating areas of circles. What differs is that in one situation a student is likely to think additively, putting together areas, and in the other situation, the student is likely to take away some areas from another. Many students find this latter type of thinking more difficult. The sizes of the circles in the two options are the same to make a common discussion easier.

No matter which option students choose, follow-up questions such as these would be appropriate:

- *What is the area of each of the full circles in your picture? How do you know?*
- *Did you need to calculate the area of each circle or semicircle to solve the problem? What did you have to do to calculate the area of the dark section?*
- *How can you test that your answer is reasonable?*

Variations. Students could be asked to create a diagram involving circles and other shapes that are colored in two colors. They then must determine the area associated with each color.

> **Option 1:** Create and solve a problem where it would be useful to know how to calculate the area of a circle.
>
> **Option 2:** Create and solve a problem where it would be useful to know how to calculate the circumference of a circle.

In both situations, students are thinking about practical applications of circle measurement. They need to recognize that situations involving circumference should involve linear units and could involve contexts such as buying fencing, walking around a circle, calculating the number of revolutions a bike wheel makes in a given distance, and so on. They need to recognize that situations involving area should involve square units and might involve contexts such as painting, putting down carpet, and so forth.

Questions applicable to both tasks include:

- *What formula for circle measurements do you need to use?*
- *What kind of situation doesn't make sense to use?*
- *What kind of situation makes sense to use? Why does it make sense?*
- *How do you know that your solution is reasonable?*

PARALLEL TASKS FOR GRADES 9–12

☀ **BIG IDEA.** The unit or tool chosen for a measurement can affect its numerical value as well as the precision of the measure.

> **Option 1:** Why does it make sense that speed is measured in a unit of the form: distance unit/time unit?
>
> **Option 2:** Why does it make sense that acceleration is measured in a unit of the form: distance unit/time unit^2?

Both of the options allow students to consider unit analysis—examining how units are appropriately combined to relate to the measurement to be taken. **Option 1** will be more comfortable for many students who have a fairly good sense of what speed measures; **Option 2** will suit students who are ready to deal with somewhat more sophisticated unit analysis.

Regardless of the option selected, follow-up questions such as these could be asked:

- *How would you describe to someone who didn't know, what it is that you are actually trying to measure?*
- *Why can't you just use a simple unit like miles, feet, pounds, and so on, to describe your measurement?*
- *Why does it make sense that your measurement unit is a fractional form?*
- *Why are both distance and time units involved?*

✳ **BIG IDEA.** The same object can be described using different measurements; sometimes the measurements are related, and other times they are independent.

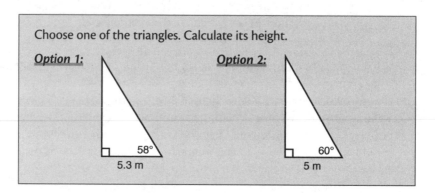

The choice of triangle allows some students to avoid more difficult decimal computations. The angles were selected to be close in value, however, to make it easier to create meaningful questions for class discussion that apply to both tasks.

Relevant questions for both tasks include:

- *How do you know the height is more than 5 m?*
- *What is your estimate for the height? Why?*
- *What trigonometric ratio would you use to help you calculate the height? Why?*
- *Could you have used a different ratio? Why or why not?*

TEACHING TIP. When creating parallel questions, it is often helpful to ensure that calculations involving both options lead to similar results. This makes it easier to ask questions related to estimating the answer.

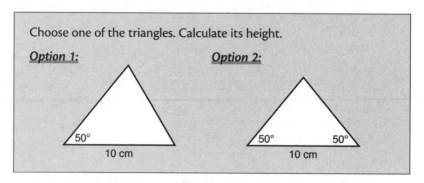

One of the triangles is symmetric. This allows students still only comfortable with simple right triangles to divide the triangle in half and solve the problem more easily than could be done with the other triangle. Whichever option is chosen, however, the students are recognizing that the height of the triangle is the common height of two internal right triangles and that it is the use of the internal right triangles that helps solve the problem.

No matter which option students choose, follow-up questions such as these would be appropriate:

- *How did you draw the height? Why did you do it that way?*
- *Were the two triangles that you formed identical?* (Note that different choices lead to different answers.)
- *Did you need to know all of the angles in the triangle to calculate the height? Explain.*
- *How did you calculate the height?*

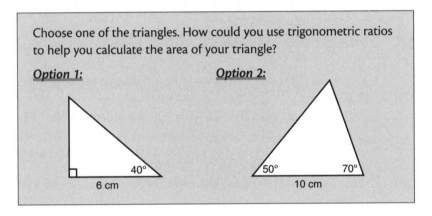

In this choice, one task involves a right triangle, which is simpler. A student who is only ready to deal with the right triangle can do so. But, in both cases, students are working with the notion that the height of a triangle can be determined by using trigonometric ratios.

Questions applicable to both tasks include:

- *What measure do you need to help you calculate the area?*
- *What trigonometric ratio will you use to help you figure out that measure? Why did you choose that ratio?*
- *What steps did you have to take to determine the area?*

Variations. Students could be asked to determine perimeters instead of areas.

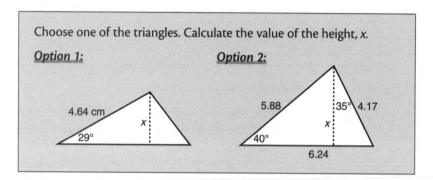

Choose one of the triangles. Calculate the value of the height, *x*.

The tasks are different in that in **Option 2** the student must choose which pieces of information to use. In **Option 1**, only two pieces of information are given and both are essential. In each case, however, students must determine which trigonometric ratio to use and why.

Regardless of the option selected, follow-up questions such as these could be asked:

- *How many of the provided pieces of information do you need to use? Explain your thinking.*
- *Could you have used other information to calculate the height? What other information could you have used? How?*
- *How could you have estimated the height? Why does that estimate make sense?*
- *What was the value for your height? How could you test to see if your answer is right?*

TEACHING TIP. One way to create parallel tasks is to provide extraneous information in one task that is not provided in the other.

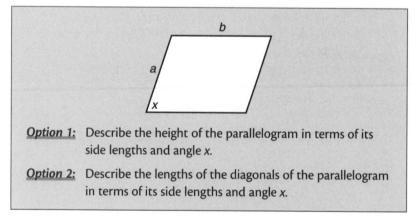

Option 1: Describe the height of the parallelogram in terms of its side lengths and angle *x*.

Option 2: Describe the lengths of the diagonals of the parallelogram in terms of its side lengths and angle *x*.

In **Option 1**, students need only use right angle trigonometry, and in **Option 2**, students need to use the law of cosines as applied to an acute triangle. In either case, however, they see that different dimensions of a parallelogram can be described in terms of the side lengths and an angle.

Even though the tasks differ in complexity, there are a number of common questions that can be asked of all students. These include:

- *When you drew the length you needed, did you create a right triangle or not?*
- *Why does it make sense that the formula you created needed to use more than one of the provided measurements?*
- *Did the formula need all three provided measurements? Why or why not?*
- *What is your formula? How could you explain to someone how you figured it out?*

✸ **BIG IDEA.** Knowing the measurements of one shape can sometimes provide information about the measurements of another shape.

> A measuring device is capable of measuring only in increments of $\frac{1}{2}$".
> Using this device, the side lengths of two different cubes are found to be $3\frac{1}{2}$".
>
> **_Option 1:_** How much wider might one cube be than the other?
>
> **_Option 2:_** How much more volume might one cube have than the other?

Issues of error of measurements are important for students to consider. They need to think about the fact that if a linear dimension is measured to the nearest $\frac{1}{2}$", then that affects the range of acceptable area measures and volume measures differently. This is a difficult concept for many students. This set of parallel tasks allows struggling students to focus on what error of measurement means, that is, in this case, that a measured length of $3\frac{1}{2}$" is actually a length between $3\frac{1}{4}$" and $3\frac{3}{4}$". It is very important that a student realize that, in fact, if two items are both reported as $3\frac{1}{2}$", one might still be longer than the other.

Students who choose will have the option of focusing on how the volume error is affected by the linear error. In this case, the volume might be anywhere between 34.3 in³ and 52.8 in³, quite a large spread. All students will benefit from realizing how error can compound.

Relevant questions for both tasks include:

- *Could the measure in one cube be one full unit more than the measure in the other?*
- *Did you need to know that the measurement was to the nearest half inch to answer the previous question?*
- *What is the least possible measure?*
- *What is the greatest possible measure?*
- *How much larger could one measurement be than the other?*

> Two cones have the same base, but the second one is 5 times as high as the first one.
>
> **_Option 1:_** How are their volumes related? Is that always true?
>
> **_Option 2:_** How are their surface areas related? Is that always true?

In either case, students might begin by creating two cones that meet the required condition. But, to generalize, students should be encouraged to use formulas. Many students will find it easier to use the volume formula required in **_Option 1_** than the surface area formula required in **_Option 2_** to determine the required relationship. It turns out that the volume relationship is independent of specific values of the radius and height, but the surface area is not.

Some common questions that suit both tasks are:

- *What formula did you use to solve the problem?*
- *What relationship do the measurements have? How did you figure that out?*
- *Do the specific values of the radius and the height affect the relationship?*

> **_Option 1:_** How could knowing the formula for the volume of a cube help you figure out the formula for the volume of a square-based pyramid?
>
> **_Option 2:_** How could knowing the formulas for the volume of a triangular pyramid and the surface area of a sphere help you figure out the formula for the volume of a sphere?

In both cases, students must visualize how one shape relates to another shape, either the pyramid inside the cube or the pyramid inside the sphere. In one case, it is assumed that students know the formula for the volume of the pyramid, and in the other case that it is something they must figure out. In **_Option 1_**, the student can think about decomposing the cube into six pyramids. In **_Option 2_**, the student can visualize the surface of the sphere as being made up of many little triangles (like a geodesic dome) which, in total, form the surface area of the sphere, with each triangle serving as the base of a pyramid and the center of the sphere as its apex.

Some common questions appropriate for both tasks include:

- *What formula did you begin with?*
- *What mental picture did you have that helped you solve the problem?*
- *How could you test if your formula makes sense?*

✹ **BIG IDEA.** **Measurement formulas allow us to rely on measurements that are simpler to access to calculate measurements that are more complex to access.**

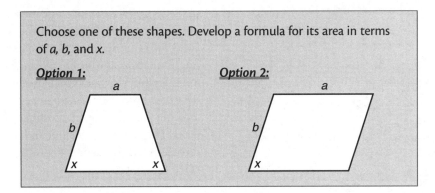

Choose one of these shapes. Develop a formula for its area in terms of *a*, *b*, and *x*.

Option 1:

Option 2:

Not only is the area formula for the isosceles trapezoid a bit more complicated for students than the area formula for the parallelogram, but students working on the trapezoid also are likely to use both sines and cosines to develop a formula.

In discussing their results for both cases, students will recognize the role of height in the area formula. In addition, students will also come to realize that different formulas are possible to describe the areas of shapes, an important measurement idea.

Even though the tasks differ in complexity, there are a number of common questions that can be asked of all students. These include:

- *What lengths do you need to know that were not directly provided?*
- *What trigonometric ratios do you need to determine those lengths? Why those ratios?*
- *Why will the formula involve a sine or cosine or tangent of x rather than the measure of x in degrees?*
- *What is your formula?*

Variations. Students could use trigonometry to create new formulas for areas of other shapes, for example, regular hexagons.

SUMMING UP

The five big ideas that underpin work in Measurement were explored in this chapter through 52 examples of open questions and parallel tasks, as well as variations of them. The instructional examples provided were designed to support differentiated instruction for students at different developmental levels, targeting two separate grade bands: grades 6–8 and grades 9–12.

Measurement is a strand that links Number with Geometry. Students experiencing weakness in either of those strands might struggle in Measurement. Because Measurement is a very practical part of the mathematics that is taught, it is important to have all students actively participate in this strand through differentiating instruction to meet students where they are developmentally.

MY OWN QUESTIONS AND TASKS

Lesson Goal: Grade Level: _____

Standard(s) Addressed:

Underlying Big Idea(s):

Open Question(s):

Parallel Tasks:

Option 1:

Option 2:

Principles to Keep in Mind:

- All open questions must allow for correct responses at a variety of levels.
- Parallel tasks need to be created with variations that allow struggling students to be successful and proficient students to be challenged.
- Questions and tasks should be constructed in such a way that will allow all students to participate together in follow-up discussions.

There are many other open questions and parallel tasks that can be created to differentiate instruction in Measurement. For example, students might explore additional trigonometric identities, they might further explore concepts of similarity and congruence, or they might explore additional alternate measurement formulas. A form such as the one shown here can serve as a convenient template for creating your own open questions and parallel tasks. The Appendix includes a full-size blank form and tips for using it to design customized teaching materials.

Data Analysis and Probability

DIFFERENTIATED LEARNING ACTIVITIES in data analysis and probability are derived from applying the NCTM process standards of problem solving, reasoning and proof, communicating, connecting, and representing to content goals of the NCTM Data Analysis and Probability Standard, including

- formulating questions that can be addressed with data, and collecting, organizing, and displaying relevant data to answer these questions
- selecting and using appropriate statistical methods to analyze data
- developing and evaluating inferences and predictions that are based on data
- understanding and applying basic concepts of probability (NCTM, 2000)

TOPICS

Before beginning the task of differentiating student learning in data analysis and probability, it is useful for teachers to have a good sense of how the topics in the strand develop over the grade bands. The NCTM *Curriculum Focal Points* (NCTM, 2006), which suggest what mathematical content should be the focus at each grade level through grade 8, were used as the basis for recommendations made in this resource for grades 6–8; the NCTM *Principles and Standards for School Mathematics* (NCTM, 2000) helped form the basis for the material for the higher grades. For a teacher at a particular grade level, it can be helpful to be aware of where students' learning is situated in relation to what learning has preceded the present grade band and what will follow.

Grades 6–8

Within this grade band, students learn to use descriptive statistics to explain central tendencies and dispersion of data sets. In particular, students learn that the appropriate measure of center—the **mean**, **median**, or **mode**—is selected based on the particular situation and purpose. They organize and display data sets using **bar graphs**, **circle graphs**, **histograms**, and **scatterplots** and use these to pose and solve problems.

In probability, students continue to organize **outcomes** of an experiment and determine **experimental probabilities**. They also learn how to calculate and use **theoretical probability** to make approximate predictions of both single and **compound events**.

Grades 9–12

Within this grade band, students build on their work in statistics from previous grades to organize and display data to pose and answer questions. Students learn to determine **quartiles** and **percentiles**, **standard deviations**, and **z-scores** to obtain more information about the location and spread of data. Students also make scatterplots and estimate **lines of best fit** and determine **regression coefficients** to make and test conjectures.

THE BIG IDEAS FOR DATA ANALYSIS AND PROBABILITY

Coherent curricula in data analysis and probability that meet NCTM content and process standards (NCTM, 2000) and support differentiated instruction can be structured around the following big ideas:

- To collect good data, it is necessary to decide what collection method is most suitable and how to best pose any questions required to collect the data.
- Visual displays quickly reveal information about data.
- Not only can information be read from visual displays, but conclusions can be drawn and inferences made.
- How data are displayed can affect what conclusions are drawn from the data.
- A summary statistic can meaningfully describe a set of data.
- Sometimes it is reasonable to generalize from a sample of collected data.
- There are a variety of ways to calculate a probability, both theoretical and experimental.
- There are a variety of representations of a **probability distribution**.

The tasks set out and the questions asked while teaching data analysis and probability should be developed to evoke these ideas. The following sections present numerous examples of application of open questions and parallel tasks in development of differentiated instruction in these big ideas across two grade bands.

OPEN QUESTIONS FOR GRADES 6–8

> **OPEN QUESTIONS** are broad-based questions that invite meaningful responses from students at many developmental levels.

✹ **BIG IDEA.** To collect good data, it is necessary to decide what collection method is most suitable and how to best pose any questions required to collect the data.

> You want to find out how students in your school feel about the new dress code. Create two survey questions: one question that may be **biased** and another question without bias. Tell why there is or is not bias.

Students may find it either too difficult or too easy if they are given an existing survey question and asked to identify the bias. In this open question, students are given the question that the surveyor wants to find the answer to, and are asked to create their own survey questions. Students might consider either the wording of the questions or the sample to which they are administered, or both. Students might start with a biased question and then modify it to make it fair, or they might start with a good question and then introduce bias of some sort. Alternatively, students might create two entirely different survey questions—one biased and one unbiased.

> The early bird catches the worm. How can we show this is true? What data would you collect and how would you organize the data?

Students need to be able to design their own experiments and decide how they would collect relevant data. This problem leaves it open to the students to choose the variables, the data they will collect, the methods they will use to collect the data, and the ways they will organize the data to draw conclusions. The statement itself is open to interpretation. Students can investigate the prediction in the literal sense, that is, measuring the number of worms a bird eats and observing how the number changes with time from sunrise onward. Alternatively, students can use an example of arriving early and reaping the benefits, for example, studying the number of sale items available at the hour a store opens versus later in the day, near closing.

If students have difficulty getting started, the teacher could ask scaffolding questions such as:

- *What is the question you want to answer?*
- *What information do you need to know in order to answer the question?*

- *Are the data you will collect numbers or not?*
- *Will you collect the data by observation or by using a survey or experiment?*
- *What do you predict will be the answer to your question? Why?*

TEACHING TIP. Sometimes two students will interpret a question quite differently. This should be regarded as a teaching opportunity and not a difficulty. Students who thought one way will be exposed to an alternate approach, even when their original idea still makes sense.

✳ **BIG IDEA.** **Visual displays quickly reveal information about data.**

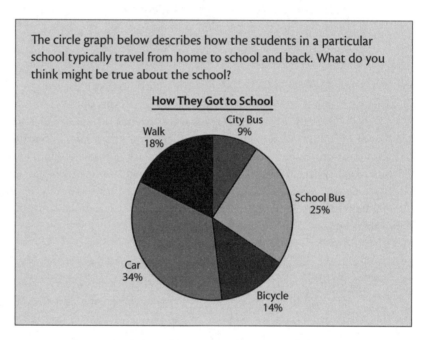

The circle graph below describes how the students in a particular school typically travel from home to school and back. What do you think might be true about the school?

How They Got to School

- City Bus 9%
- Walk 18%
- School Bus 25%
- Car 34%
- Bicycle 14%

Students can choose to talk about all or part of the graph; this flexibility may make the question more accessible to some students. They might make inferences about how far away students described in the graph live from school, or they might consider why the various non-walker percentages split the way they do. Some students might try to guess whether the school is large or small and support their thinking with the data. Class discussions should emphasize the value of a visual display and highlight how much information can be drawn from the data.

TEACHING TIP. Sharing of student work and strategies should be encouraged, especially when students are responding to questions that give them the opportunity to be creative in their answers.

✳ **BIG IDEA.** Not only can information be read from visual displays, but conclusions can be drawn and inferences made.

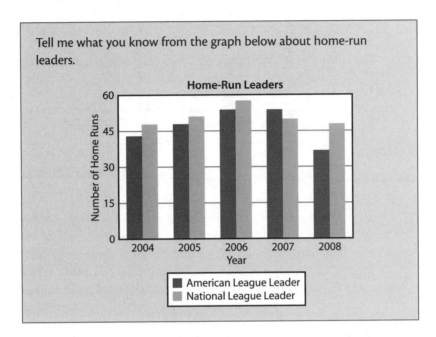

Tell me what you know from the graph below about home-run leaders.

This open question allows all students to actively participate in the discussion about information that can be read or inferred from the given graph. By presenting data graphically on a topic that is likely to be of natural interest to many students, the real-world value of the seemingly academic pursuit of data analysis is brought home. Students will have lots of interesting things to say about the home-run leaders in the two leagues from 2004 to 2008. For example, some will be happy to point out that, except for the next to last year, the National League had more home runs than the American League. Other students might look at trends over the 5 years and even try to account for the changes in home-run leaders.

For students who do not know how to begin, questions that could be asked include:

- *What kind of graph is being used to show the home-run leaders?*
- *What information about home runs is shown in the graph?*
- *How does the number of home runs by the American league leader compare to the number by the National league leader?*
- *What kind of trends are you seeing over the time period shown?*

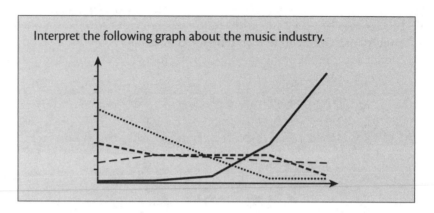

Interpret the following graph about the music industry.

Interpreting graphs involves telling stories, something most students enjoy. By not providing any labels or titles but still telling students that the graph has to do with the music industry, they are given a starting point and a lot of freedom in how to proceed. Some students might need a few more suggestions, for example, indicating that they might think about CDs, song downloads, sale of MP3 players, or perhaps musicians who play different types of instruments. Most students are accustomed to being provided very specific graphs to address very specific questions and do not have the opportunity to think more deeply about data trends. This open question provides them the opportunity to not only read graphs but also to make inferences based on the information presented in a graph.

Variations. Similar analysis opportunities could be offered by using other types of graphs—such as circle graphs, bar graphs, or histograms—or a different subject context (e.g., a focus on Internet use rather than the music industry).

A **pictograph** about recycling uses the key shown at the right.

What do you think the data might be like to make this key appropriate?

= 10

Rather than asking students to create a pictograph, this question encourages them to realize that there are many possible ways of using a key to describe what a symbol represents for a graph. Students might suggest that the recycling symbol represents 10 recycling bins, 10 units of time, or 10 reusable shopping bags. They might also think about the values in the data set that would make a scale of 10:1 appropriate. They might suggest that this key would be best suited to values ranging from about 10 to 100.

Variations. An alternate key symbol (suggesting alternate topics) or a different value for the symbol could be proposed.

TEACHING TIP. One way to open up questions is to provide only a minimal amount of information, leaving more choices to the student for how to use the data.

✳ BIG IDEA. **How data are displayed can affect what conclusions are drawn from the data.**

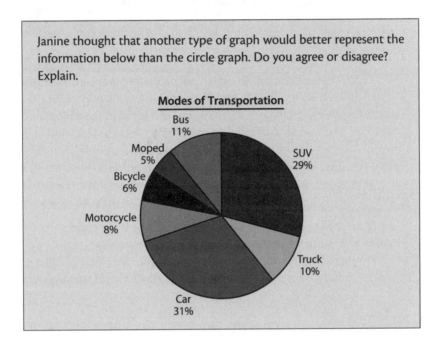

Janine thought that another type of graph would better represent the information below than the circle graph. Do you agree or disagree? Explain.

Modes of Transportation

- Bus 11%
- Moped 5%
- Bicycle 6%
- Motorcycle 8%
- SUV 29%
- Truck 10%
- Car 31%

This question allows students to discuss the merits of a circle graph and when its use is appropriate. Some students will think that a bar graph or line graph would be a better representation of the information in this case because the circle graph is crowded with too many attributes. Others will find that the circle graph is a clear way to show the percentages assigned to each mode of transportation.

Students can be prompted to think about what information is missing from this type of display (e.g., what the percentages actually represent, the number of data points each wedge represents, the "stability" of the data in a larger **population**). The relative importance of the information that is *not* presented could affect the decision about what type of graph is best in any given circumstance.

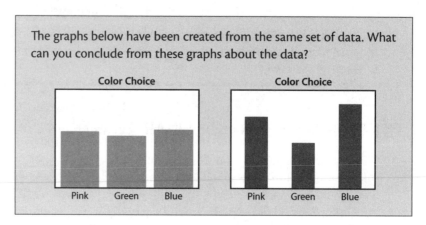

The graphs below have been created from the same set of data. What can you conclude from these graphs about the data?

Color Choice / Pink Green Blue

Color Choice / Pink Green Blue

Often students are presented with a misleading graph and are asked to describe *why* it is misleading. In this case, although students may realize that one of the graphs is misleading, they have to make the decision as to which one is misleading and why. The openness of this approach empowers students and encourages individual decision-making. This type of question also allows the students to compare the two different graphs and decide what each one tells them about the data.

If students struggle to come up with a response, the teacher could provide scaffolding with questions such as these:

- *What do the bars on the graphs represent about the colors?*
- *Does it matter that the y-axis value labels are missing?*
- *How can both graphs represent the same data and look so different?*
- *What is different about the two graphs? Which would you use when?*

✳ BIG IDEA. **A summary statistic can meaningfully describe a set of data.**

The following numbers (in any order) represent the mean, median, mode, and **range** for a set of data:

13, 14, 15, 8

What could this data set be?

This question is open because students can choose which number is the mean, which the median, which the mode, and which the range from the given group of numbers. They can also come up with their own set of data—either a fairly simple set or a much more complex set—that will work for their choices. There will be a variety of answers to this question.

For students having difficulty getting started, questions such as these could be helpful:

- *Which number would make the most sense for the range? Why?*
- *Does the value you chose for the mean have to be one of the data values? How about the median? The mode?*

> You take a part-time job with a company that has 10 employees, and the company tells you that the average salary is $200 per day. What do you expect your salary to be?

This situation leaves a lot of room for students to make decisions. The salary of the employees could be $100 each for all but one of them, who may make $1,100 per day; it could be that some make $200 per day, some $100 per day, and some $300 per day; it could be that everyone makes $200 per day. The latter situation will be simplest for some students, but the wide range of possibilities makes the question engaging for students of all abilities.

Discussion will likely occur around the unknowns, such as the range of salaries, how many hours is considered a day's work, and whether the employees are paid a daily rate or an hourly rate.

> Create a sentence that makes mathematical sense and that uses these words and numbers:
>
> *median, mean, more, 20*

This open question allows students to demonstrate what they know about the concepts of mean and median in a fairly creative way. Students might offer any of these sentences, for example:

- *A median can be 20 more than a mean if there are some really low data values.*
- *The mean and median of the data set 100, 200, 300, 400, are both more than 20.*
- *The mean is 20 more than the median if the data set is 10, 20, 90.*

Variations. Different sets of words and numbers could be used. Some options include: *range, mean, 20, 10; median, average, data, 3; range, spread, 8, 28.*

✳ **BIG IDEA.** There are a variety of ways to calculate a probability, both theoretical and experimental.

> How many gray socks and how many white socks are in a drawer if the probability of randomly choosing two socks, one of each color, is close to $\frac{1}{3}$?

This question is opened up to students who can imagine a variety of scenarios with different numbers of gray and white socks in the drawer to meet the conditions of the problem. Students will realize they have to start with at least three socks. Then they can build on that to decide how many of each color can be in the drawer and how many tries are needed to pull out one of each color to end up with a probability close to $\frac{1}{3}$. Not stating exactly what the probability is allows for more possibilities.

If students do not know how to begin, the teacher could ask scaffolding questions such as:

- *Could there be 1 gray sock and 1 white sock in the drawer?*
- *Could there be 2 gray socks and 2 white socks?*
- *Do the numbers of gray socks and white socks have to be the same?*
- *Does the probability depend on whether there are also socks of other colors in the drawer?*
- *How would you calculate the probability of pulling out two socks of the same color?*
- *What do you need to know to calculate the theoretical probability of pulling out two socks of different colors?*

> Yvonne said that the probability of two times in one hour ending in the same last digit—such as 2:12 and 2:22—is very small. Do you agree with her?

One of the things that makes this question open is the use of the phrase "very small." With the question being somewhat vague, students have more latitude in answering; there are no rules about what very small means. In this case, some students are likely to think that the probability of the **event** is $\frac{1}{10}$, recognizing that there are 10 possible final digits. Others will start to look at the possible outcomes 2:02, 2:12, 2:22, and so forth, and realize that using two different times means reducing the outcomes by one, and thus the total number of times to pick from is 60 less the one that you cannot choose again. As students work through the possibilities, many ideas around calculating theoretical probability will surface through discussion of this one question.

TEACHING TIP. Questions that allow students to choose to agree or disagree are useful for many students because the question gives them a starting point. Rather than having to originate an entire response, all they have to do is come up with a reason for their choice.

OPEN QUESTIONS FOR GRADES 9–12

✳ **BIG IDEA.** To collect good data, it is necessary to decide what collection method is most suitable and how to best pose any questions required to collect the data.

> A survey was carried out using a **stratified sampling** method to find out about customer satisfaction in a department store. How might the sample have been stratified?

This question is open because students can choose the design of the stratified sample. If students are uncertain about the meaning of "stratified," it is perfectly acceptable to clarify, for example, by indicating that there are subgroups of customers and that they are represented in the sample according to the proportion of total customers in each subgroup. Students can decide which characteristics they would consider when setting up their subgroups of the **sample population** and why a store might be interested in those subgroups. The context is one that all students are familiar with, and this will help them focus on the importance of the sampling method to ensure the accuracy of the results.

> A researcher is investigating a target population that consists of males, ages 18–25, who have had their first car accident resulting in damage costs. What question(s) might the researcher pose?

It is useful for students to think about what questions might be asked of or about this particular target population. Students may focus on why the researcher would be interested in this group, on the kinds of questions a researcher would ask about this group, or on the questions a researcher would ask of members of this group in a survey. No matter how they focus, students are thinking about the collection of data to support a **hypothesis**.

For students who struggle to come up with a response, questions that could be asked include:

- *Why would the researcher want to speak to only males?*
- *What questions would you have around young drivers and car accidents?*
- *Who would be interested in surveying this target population?*
- *What could the researcher be interested in finding out about this target group?*
- *Can you think of a survey question that you would ask them?*

✳ **BIG IDEA.** **Visual displays quickly reveal information about data.**

> Create two graphs that show:
>
> (a) the same kind of data but at different times or places
>
> (b) different kinds of data at the same time or place

Instead of providing two graphs and having students compare the data, this question has students creating the graphs themselves. The question does not ask for any particular type of graph, thus allowing the students to choose which type of graph is best for comparing the two sets of data they have chosen. For example, bar graphs, double bar graphs, histograms, circle graphs, and so forth, might be chosen, depending on the complexity of the data and type of message students want to convey. The question is open to students at all levels because they are able to pick numbers and the type of graph they are most comfortable with.

TEACHING TIP. By providing only a premise for the students to work with and not actual numbers, the question becomes accessible to a broader range of students.

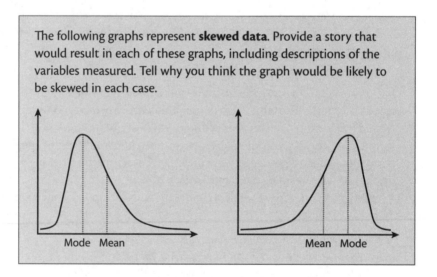

> The following graphs represent **skewed data**. Provide a story that would result in each of these graphs, including descriptions of the variables measured. Tell why you think the graph would be likely to be skewed in each case.
>
> Mode Mean   Mean Mode

Students are often shown graphs of skewed data and told where the mean and mode are, but they are not given the opportunity to imagine what data sets would actually produce a graph that looked like the graphs given. The question is open in asking students to interpret what each graph is telling them and in asking them to find real-life examples of skewed data.

If students have difficulty getting started, the teacher could provide scaffolding with questions such as these:

- *Can you think of a set of data where the mean is greater than the mode? Where the mode is greater than the mean?*
- *How can one extreme value affect the mean and mode? How can it affect the whole graph?*
- *Which graph would more likely represent salaries in a company? Can you think of a different set of data that would result in a similar graph?*

✴ **BIG IDEA.** **Not only can information be read from visual displays, but conclusions can be drawn and inferences made.**

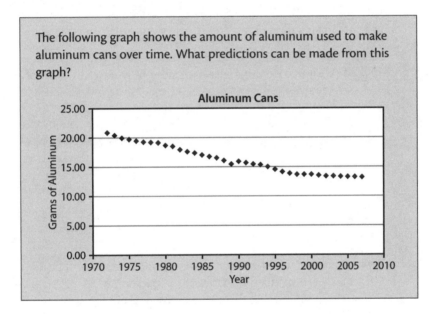

The following graph shows the amount of aluminum used to make aluminum cans over time. What predictions can be made from this graph?

In not specifying what types of predictions students should make about the mass of aluminum cans, this question is open. Some students might try to draw a line of best fit and use that to **extrapolate** to later years the amount of aluminum used to make cans in the future. Others might make predictions based on trends, noting particularly the change over the last 10 years.

For students who do not know how to begin, questions such as these could be helpful:

- *What does the graph tell you about the mass of aluminum cans?*
- *What trends do you see in the graph?*
- *Could you extrapolate to predict what may happen in a later year?*
- *How sure can you be about your extrapolations?*

✴ BIG IDEA. **How data are displayed can affect what conclusions are drawn from the data.**

> A marketing company wants to use a graph to show the increase in customers switching from one phone company to another. What could this graph look like?

The question is open in allowing students to choose any type of graph to show that more customers switched from one phone company to another over time. Many aspects of the graph can be manipulated. These include the scale of the graph, the time factor, the type of graph, and what is actually being graphed (e.g., percentage changes, numbers of customers, etc.). Students will come up with different representations in their graphical displays to convince others of their findings.

✴ BIG IDEA. **A summary statistic can meaningfully describe a set of data.**

> The score on Day 8 is fairly typical of the rest of the scores. What can you expect on Day 8? How many standard deviations is this score away from the mean?
>
> | Day | 1 | 2 | 3 | 4 | 5 | 6 | 7 | 8 |
> |-------|-----|-----|-----|-----|-----|-----|-----|-----|
> | Score | 49 | 44 | 37 | 45 | 45 | 42 | 48 | ☐ |

Many data management questions provide data sets and have students calculate the mean and standard deviation. This question provides an incomplete set of scores and a condition that the missing score is typical. A student is likely to interpret this as meaning that the score is not far from the mean. Students can decide on a score for the 8th day and then determine how many standard deviations away from the mean that score is. This will help students understand the relationship between the mean and the standard deviation of a set of data.

If students struggle to come up with a response, the teacher could ask scaffolding questions such as:

- *Do you think 35 is typical of the rest of the scores? Why or why not?*
- *What about 40?*
- *Can you find the mean?*
- *What information do you need to calculate the mean?*
- *What information do you need to calculate the standard deviation?*

Variations. Rather than being asked to choose a score that is typical, students could be asked to choose a score that is not typical or is an outlier to see its effect on the mean and standard deviation of the set of scores.

> Describe the similarities and differences between the graphs of two
> **normally distributed** sets of data:
>
> (a) a data set with mean of 12 and standard deviation of 6
>
> (b) a data set with mean of 12 and standard deviation of 2

To compare the two graphs, students can start with what they know: that the two data sets have the same mean and different standard deviations. Some students might create actual sets of data or graphs to help them visualize the two situations. Although the means are the same and both graphs are bell-shaped, one graph will be wider and shorter whereas the other will be taller and narrower. Students will come to realize that normally distributed data are described by both their mean and their standard deviation.

Variations. Instead of the same mean and different standard deviation, the two graphs could have the same standard deviation and different means.

✴ BIG IDEA. **Sometimes it is reasonable to generalize from a sample of collected data.**

> Austin and Zachary are in different classes. Austin's math mark was
> 89%; Zachary's was 88%. Can you tell who is better at math? How
> would you know?

Statistical analysis can be used not only to look at a sample of collected data but also to examine and compare two sets of data. This question is open because students will make their own decisions about how they will back up their conclusion. Some students will argue that 89% is higher than 88% and thus Austin is better at math. Other students will start to ask questions about how the mark was attained in each of the math classes (e.g., it might be affected by tests, assignments, projects, or even attendance). Class discussion should ultimately lead students to understand that they need to standardize the classroom data to meaningfully compare the marks.

✳ **BIG IDEA.** There are a variety of ways to calculate a probability, both theoretical and experimental.

> Cathy and her friend have been buying lottery tickets weekly for the last 5 years. Cathy has used the same five numbers every time, while Sonya believes in changing her numbers every week. Who has the better strategy to increase her chances of winning in the next drawing? Explain your reasoning.

In the study of data management, open questions that allow students to have a rich discussion around "myths" or common misconceptions are always engaging for them. This question draws out theoretical probabilities of winning, but framed in a way that allows all students to compare the two situations set up for buying the tickets.

For students having difficulty getting started, questions that could be asked include:

- *Which strategy would you choose? Why?*
- *Suppose there are two drawings in a row one night. Are they **dependent** or **independent events**?*
- *How would you calculate the theoretical probability of winning a lottery?*
- *What information about the lottery do you need to do this?*

> Toss a pair of dice, one red and one blue. Consider these events:
>
> - Event A is "sum of dice is 6"
> - Event B is "blue die shows a 4"
>
> Tell what you know about the probabilities of these two events occurring.

Asking students to tell what they know is always an open approach. If they wish, students can make a visual chart of the red and blue dice to help them see the events more clearly. Some students might simply look at the probabilities of each event separately, while other students might venture into the probabilities of compound events.

The students are free to consider either theoretical or experimental probabilities. Some students might engage in an experiment to see how often a sum of 6 occurs, or how often the blue die shows a 4, or both at the same time.

Variations. A possible variation on this question could be to ask what pair of events would satisfy the following set of conditions: Event A has to relate to the sum of the values on the dice. Event B has to relate to the difference of the values on the dice. Event B must be much more likely than Event A.

The following is a **frequency distribution graph** of an experiment with two dice. Meghan said that if she rolls two dice again, the total will not be a 9 and most likely will be a 7.

Do you agree with her? Explain why or why not.

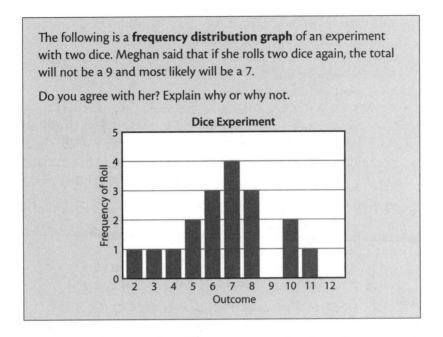

By asking students whether they agree or not, they have the opportunity to express their ideas around experimental and theoretical probability. Some students will agree because the results of the experiment to that point could lead to this type of prediction. Other students will argue that knowing that the theoretical probability of rolling a sum of 9 is not 0 means that there *is* a possibility of showing a 9 in the next trial. A discussion can take place around the likelihood of showing a 7 with the two dice—based on the experimental results and what students may know about the theoretical probability of showing a 7.

If students do not know how to begin, the teacher could provide scaffolding with questions such as these:

- *What information does the frequency distribution graph tell you about the outcomes of the experiment?*
- *What predictions can be made based on what you know from the graph?*
- *If you carried on with the dice experiment for another 10 rolls, what do you think the graph would look like? Another 10 rolls?*

At the school dance, there is a drawing for which 500 raffle tickets are sold and only one winner will be drawn. You and your friend are going to buy tickets. How many tickets do you and your friend need to buy to give one or the other of you a good chance at winning?

This type of open question has students discussing what a "good chance" means and who can do the winning. If students decide that the two individuals winning are **mutually exclusive events** and that both cannot be winners, they will have to figure out how many tickets each should buy to give either one a higher probability of winning. Some students may choose to have one person buy zero tickets and the friend buy many tickets. A rich discussion on how many tickets it would take to give that "good chance" could ensue: Is it 75%, 80%, or 99% of the total tickets?

Alternatively, some students might regard the lottery ticket purchase as a "team" activity, where the objective is simply to have one member of the team win the drawing, with the intent of sharing the winnings. In this case, the discussion might focus not only on the total number of tickets to be purchased but also on each team member's contribution and on distribution of the potential winnings.

Variations. The problem could be altered by varying the number of winners in the draw. For example, three raffle tickets might be drawn for prizes of equal value.

> A solution to a problem is $_{12}P_4$. What was the problem?
>
> Modify your problem to give a solution of $_{12}C_4$.

When working with permutations and combinations, students need to understand how they are related to each other. Letting students choose the context opens up this **combinatorics** question and encourages thinking around what the differences are between permutations and combinations. Students do not find it as challenging when faced with a question that is clearly either a permutation or a combination problem and they can simply complete the calculation by hand or using a calculator. When students must come up with their own problem, they think beyond simple calculations and focus more on the meaning of permutation or combination.

> A 5-card hand is dealt from a deck of 52 cards. Some hands are worth more points than others in card games. Using your knowledge of probabilities, explain why that makes sense.

Understanding the calculations for card hands dealt from a regular deck of cards can be complex because of the number of variations involving suits, face cards, and values. This question is open because, although all students are working with 5-card hands, some students can begin by looking at just one pair versus two pairs or at hands made up of all cards of one suit. Other students might start by addressing the probability of dealing a hand that includes three of a kind and a pair, or four of a kind, and realizing why such hands are worth more. In any case, students will be looking at ways of combining 5 cards chosen from 52 cards and will be able to offer explanations based on mathematical principles.

✹ **BIG IDEA.** There are a variety of representations of a probability distribution.

The rules of a spinner game:

        Red—win 3 points

        Blue—win 4 points

        Yellow—win 5 points

        Green—lose 3 points

        Purple—lose 5 points

Draw a spinner with five separate regions where the **expected value** is to lose 1 point.

In the question, a description of a game is provided with points for landing in certain regions. The question is open because students can design their spinner in any way they wish, as long as the expected value works out to losing 1 point. Instead of providing a spinner divided into predetermined regions, the question essentially starts with the "answer" and has the student making the decisions on how to arrive at that answer. Various solutions exist, and the different spinners created can be compared among members of the class.

Variations. Instead of a spinner, blank dice could be used or a simple board game could be designed for gaining and losing points. In the new setting, either an expected value could be provided or students could be asked to design a "fair" game where the expected value is zero.

TEACHING TIP. A good strategy for opening up questions is to start with the answer. This allows for a variety of responses from students as they work out questions that will produce the specified answer.

The following problem is presented:

    A coin is tossed three times.

    Find the probability of getting exactly two heads.

Janice says that the problem involves a **binomial distribution** because there are two outcomes—heads or tails. Tim says that's not the case because there are three outcomes, since the coin is tossed three times. Who do you think is correct and why?

When students are presented with problems involving a binomial distribution, they quickly turn to using a formula to find the likelihood of a random sample having a certain number of outcomes. This coin problem is not difficult and can be solved also by using a **tree diagram** and examining the **sample space**. Here students have to consider whether the problem describes a binomial distribution or not, depending on with whom they agree. Some students will recognize that there are only two outcomes for each trial—heads or tails—which is a property of a binomial distribution. Other students will go further and recognize that they are looking at independent trials. Students who agree with Tim and focus on the three trials might work through the solution using the sample space and discover that using the formula for finding probabilities of an outcome for a binomial distribution produces the same answer.

PARALLEL TASKS FOR GRADES 6–8

PARALLEL TASKS are sets of two or more related tasks that explore the same big idea but are designed to suit the needs of students at different developmental levels. The tasks are similar enough in context that all students can participate fully in a single follow-up discussion.

✳ BIG IDEA. To collect good data, it is necessary to decide what collection method is most suitable and how to best pose any questions required to collect the data.

Option 1: You want to conduct a survey to help you decide what two flavors of ice cream to order for an upcoming school event.

    (a) Create a biased question for the survey. Rewrite the question so that it is unbiased.

    (b) How will you collect the data?

Option 2: You want to conduct a survey to collect reliable data for stocking an ice cream stand with the most popular flavors.

    (a) Create a biased question for the survey. Rewrite the question so that it is unbiased.

    (b) How will you collect the data?

Although both options are about collecting reliable data from a survey, _**Option 2**_ is deliberately more vague than _**Option 1**_. The choice between the two options is not necessarily about difficulty but about the readiness or willingness of the student to tackle a question that is less direct in its instructions. The openness of _**Option 2**_ may not make it appear to be more difficult for students if they instead see it as an opportunity to set up their own surveys for the situation.

No matter which option students choose, follow-up questions such as these would be appropriate:

- *Who will you survey?*
- *How will you collect the data from your survey?*
- *What made your question biased? Were you trying to influence the results in some way?*
- *How did you change your question so that is was no longer biased?*

Variations. Students could be given a biased question and asked to make it unbiased. Students could be asked to consider situations in which a biased (or unbiased) question might be likely. Students could be given a similar task about a topic other than ice cream.

✳ **BIG IDEA.** **Visual displays quickly reveal information about data.**

The tables below display information about which types of movies are rented by boys and girls in one day. Choose one table and, using the information provided, create a graph or graphs to display the data.

Option 1:

Movie Type	Number Rented	
	Girls	Boys
Comedy	28	23
Romance	19	8
Drama	15	17
Action	11	24

Option 2:

Age	Frequency of Rentals			
	Girls		Boys	
	Comedy/ Romance	Action/ Drama	Comedy/ Romance	Action/ Drama
9–11	21	2	14	10
12–14	13	5	8	9
15–17	8	11	2	11
18–20	5	8	7	11

In *Option 1*, students can choose whatever type of graph that they feel provides the best visual display of the data from the table. This could be a double bar graph or two circle graphs—or even a pictograph—to compare the boys' and girls' results.

Option 2 uses the same data set, but groups movie types and shows how the data are distributed by age ranges. As for *Option 1*, students can choose the type of graph they want to create; in this case, a histogram might work well. Some students will find that including intervals in their graphs increases the degree of complexity.

Questions applicable to both tasks include:

- *How did you decide on which type of graph to use?*
- *What steps did you take to create your graph?*
- *What kind of information can you see clearly from your graph?*
- *What differences, if any, do you see between the choices of girls versus boys?*
- *When would you use the table and when would you use the graph?*

TEACHING TIP. Sometimes changing the complexity of the information provided is a quick way to create parallel tasks.

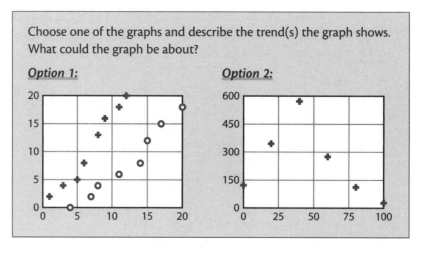

Choose one of the graphs and describe the trend(s) the graph shows. What could the graph be about?

Option 1:   *Option 2:*

Both options require students to create their own scenarios to fit the graphs. Notice that the graphs do not have any titles, which makes the task even more open. *Option 1* has smaller numbers and shows two different sets of data, both displaying positive trends. This type of trend is more intuitive for students, who might think of an example such as the more money one has the more games one can play in an arcade. *Option 2* shows larger numbers and a trend that shows an upward climb to a maximum point and then a decrease. These factors may make this option more challenging for some students.

Regardless of the option selected, follow-up questions such as these could be asked:

- *What is the trend you see in your data?*
- *How would you explain why that particular trend occurs?*
- *What might your graph be about?*
- *What could the titles on your axes be?*
- *What might your graph look like if you added more data?*

The growth of a plant over time is monitored by counting the leaves. The data are recorded in the following table:

Day	1	3	6	7	8	9	10
Number of Leaves	0	1	3	4	6	9	13

Option 1: Based on these results, how many leaves do you think the plant will have after 15 days?

Option 2: Based on these results, how many leaves do you think the plant had after 5 days?

This parallel task asks students to make predictions from the range of data given about the growth of a plant over time. For both options, students can use the table to make a graph of days planted versus number of leaves counted. *Option 1* has students making predictions by **extrapolating** and extending the graph beyond the 10 days given. *Option 2* has students using **interpolation** to find the unknown number of leaves after 5 days. Some students will choose *Option 2* because estimating a point between Day 3 and Day 6 is easier than extending the line past 10 days, since the relationship between growth and time is clearly not linear.

Relevant questions for both tasks include:

- *Did you use a graph to display the data? What kind of graph did you make?*
- *How would you describe the growth of the plant?*
- *Did it grow the same amount each day? Is that reasonable?*
- *What was challenging in finding the number of leaves for the plant for our chosen number of days after planting?*

✹ BIG IDEA. **How data are displayed can affect what conclusions are drawn from the data.**

Option 1: The profits of a small widgets company over the year are shown in the table below. Make a graph to deliberately show an exaggerated increase in profit during the summer months.

Month	Jan	Feb	Mar	Apr	May	Jun
Profit in $	3,930	5,290	5,430	6,500	6,830	5,450

Month	Jul	Aug	Sep	Oct	Nov	Dec
Profit in $	7,470	6,570	4,350	3,400	4,320	3,890

Option 2: The profit of a widgets company is $346,000 in its first year, $412,000 in its second year, $568,000 in its third year, and $602,000 in its fourth year. Make a graph to deliberately show an exaggerated increase in profit over the four years.

Both options ask students to create a misleading graph. Some students will find it easier to work with the table of monthly profits and exaggerate the results by isolating just the summer months or adjusting the size of the vertical scale. Other students will prefer to work with just four pieces of data for the company and change the scale to dramatize the increase in profits. For both options, the type of graph is chosen by the student.

No matter which option students choose, follow-up questions such as these would be appropriate:

- *How would you describe the profits of your company?*
- *What kind of graph did you use to display your profits? Why?*
- *How did you exaggerate the profits of your company?*

✸ **BIG IDEA.** **A summary statistic can meaningfully describe a set of data.**

> There are seven data points describing the number of red lights Jeremy came across when driving to work each day in a week.
>
> **_Option 1:_** The mean and the median are the same. What could the data points be?
>
> **_Option 2:_** The mode is less than the median, and the median is less than the mean. What could the data points be?

For **_Option 1_**, students must find data points such that the middle number of the data set is the same as the mean for the set. They can stack linking cubes or counters for each of the seven days as follows:

In this simplest case, the mean is 5 and the median is also 5.

Some students may stop there, which would be enough to satisfy the criteria in **_Option 1_**. However, the teacher could prompt those students by asking whether this situation would be likely to happen in real life. To create a more realistic situation, the student might move around the counters to produce something like this, where the median is still equal to the mean:

Option 2 is a more challenging task because the criteria involve all three measures of central tendency: median, mode, and mean. The student would have to manipulate the numbers to meet the given requirements of the sizes of the three types of averages for the data set.

Questions applicable to both tasks include:

- *How did you decide on the number of red lights per day?*
- *Did you choose the median or the mean first? Why?*
- *How could you use a model to help you solve the problem?*
- *How would your model show the mode? The mean? The median?*
- *Would the task have been easier or more difficult with an even number of data points?*

In order to win an award, Liam had to bring the average mark for his six courses from 78% to 80% before the end of term.

Option 1:   Liam reached his goal of 80% by improving one of his six marks. By how much did he increase his mark in the one course? How did this affect the range of his marks?

Option 2:   Liam reached his goal of 80% by improving four of his six marks. By how much did he increase his marks in each of these courses? How did this affect the range of marks?

Both options require the students to think about the effect of one or more data points on the final mean of the set. *Option 1* involves a change in only one of the six marks, but *Option 2* requires students to think about changes in four marks to produce the same change in the mean.

Regardless of the option selected, follow-up questions such as these could be asked:

- *Why don't you have to change the marks of all six courses?*
- *How do you know that at least some of the new marks were more than 80%?*
- *Can you just increase the marks in each of the courses Liam was in by 2%?*
- *What effect did the increase have on the range?*
- *How did the increase affect the median? Did it also change by 2%?*

✳ **BIG IDEA.** There are a variety of ways to calculate a probability, both theoretical and experimental.

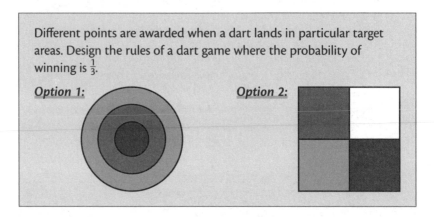

Different points are awarded when a dart lands in particular target areas. Design the rules of a dart game where the probability of winning is $\frac{1}{3}$.

Option 1:   *Option 2:*

This parallel task gives the student two choices of targets to work with. Either option requires students to use their knowledge of calculating theoretical probabilities. Students working with the more traditional-looking target for darts would need to work with areas of circles, while students choosing the square target would work with four identical areas. Both options allow students to devise a game as simple or as complex as they want.

Relevant questions for both tasks include:

- *Should it be difficult or easy to win? How do you know?*
- *How did you determine all the possible outcomes for your dart game?*
- *Did you need to consider measurements of parts of the target? Why?*
- *How did you calculate those measurements?*
- *How did you calculate the probability of those outcomes?*
- *If you were to play your game five times, how many times do you think you would win?*

PARALLEL TASKS FOR GRADES 9–12

✳ **BIG IDEA.** To collect good data, it is necessary to decide what collection method is most suitable and how to best pose any questions required to collect the data.

Write a plan to explain how you would carry out a survey for one of the following:

Option 1: You want to find out the favorite computer games played by students from your school.

Option 2: You want to investigate the favorite computer games played by youth in North America.

Both options allow students to make a plan for gathering data to answer a question. In each case, some sort of survey would be conducted. ***Option 1*** requires working with the population of students in the student's own school. ***Option 2*** has students answering a question about a population on a much larger scale—North American youth. In both cases, the student must make decisions about what sample he or she would use, the best collection method for that sample, and even the type of survey questions that would be best suited to answering the question.

No matter which option students choose, follow-up questions such as these would be appropriate:

- *What do you predict to be the favorite computer game?*
- *What is your sample population and how did you decide on it?*
- *Which sampling method would you use to collect your survey data?*
- *How would you ensure that your survey is not biased?*
- *What type of graph would you use to represent the data you collected? Why that form?*

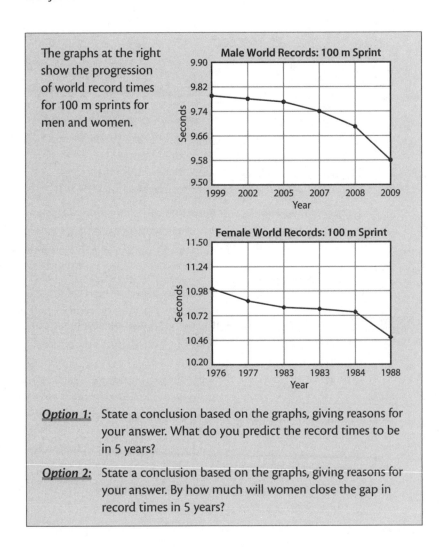

The graphs at the right show the progression of world record times for 100 m sprints for men and women.

Male World Records: 100 m Sprint

Female World Records: 100 m Sprint

Option 1: State a conclusion based on the graphs, giving reasons for your answer. What do you predict the record times to be in 5 years?

Option 2: State a conclusion based on the graphs, giving reasons for your answer. By how much will women close the gap in record times in 5 years?

Students are asked to draw conclusions based on the information displayed in the two graphs. They might discuss why lines might have been used even though the data are actually **discrete** (to make the trends clearer). *Option 1* asks students to make predictions for future years based on the information within each graph. *Option 2* asks students to consider both graphs to help them decide how much women's times will catch up with men's times in 5 years.

Questions applicable to both tasks include:

* *What information did you gather from the graphs provided?*
* *What is misleading about the way the intervals are displayed in these graphs?*
* *Was there enough information on these graphs to be confident in your predictions?*
* *What are your predictions? Explain why they make sense.*

✳ BIG IDEA. **How data are displayed can affect what conclusions are drawn from the data.**

The following table shows the results for a restaurant server skills test given to new employees and their customer satisfaction ratings.

Test Score	98	57	82	76	65	91	84	82	85	68	78	70	68	75	98	93
Customer Rating	95	81	83	90	70	65	55	67	83	98	88	60	94	78	63	70

Option 1: Use the data to show that the skills test is a good indicator of the customer satisfaction rating.

Option 2: Use the data to show that the skills test is not a good indicator of the customer satisfaction rating.

The options are designed to have students think about how to display the provided data in a way that will convince the reader either that the skills test should be used to hire new employees or that it is not useful to predict whether customers will appreciate the new employee. Some students will find it easier to use visual displays to show that the test is a good indicator, and other students will find it easier to show that it is not.

Regardless of the option selected, follow-up questions such as these could be asked:

* *Did you have to consider all of the data or just most of them?*
* *Could you use a graph to help you decide whether the test is useful?*
* *How did you show that the skills test was useful or not useful?*
* *Which combinations of scores are most important to your argument—high scores on both, low scores on both, or one high score and one low score?*
* *Is your analysis of the data valid? Why or why not?*

✳ **BIG IDEA.** A summary statistic can meaningfully describe a set of data.

> Create a set of data with eight values meeting the stated condition.
>
> **_Option 1:_** The mean and median values are the same.
>
> **_Option 2:_** The mode and the median are less than the mean.

The two tasks require students to create a set of data with choice as to the criteria the data set must meet. In **_Option 1_**, students consider only two measures of central tendency, whereas **_Option 2_** involves the mean, median, and mode. High school students continue the work from earlier grades with measures of central tendency; however, the emphasis shifts to understanding the relationships between these measures and how changes in data sets can affect the measures.

Even though the tasks differ, there are common questions that could be asked of all students. These include:

- *Could all the values be the same?*
- *Might there be some really low values but no really high ones?*
- *What strategy did you use to come up with your data set?*
- *Are your numbers close together or spread out? Did they have to be?*

> The **coefficient of determination** (r^2) for the following data set is close to 1.
>
> | x | y |
> |-----|-----|
> | 2.5 | 6 |
> | 4.8 | 4.5 |
> | 7.8 | 2.5 |
> | 5.4 | 4.1 |
> | 4.1 | 4.9 |
> | 1.5 | 6.7 |
> | 3.6 | 5.3 |
>
> **_Option 1:_** Change two points to make r^2 fall between 0.7 and 0.9.
>
> **_Option 2:_** Change four points to make r^2 fall between 0.7 and 0.9 using nonlinear regression.

Students are likely to wonder initially whether changing two points or four points is an easier way to change the value of r^2. Students will wonder whether they should change the two or four values in the same or opposite directions, or whether they should change the values a little bit or dramatically. For either option, however, they see effects of changing data points on how well the data fit the line or curve of best fit.

Relevant questions for both tasks include:

- *How did you know which point or points to change to get to your desired coefficient of determination?*
- *What calculations did you have to make in order to solve your problem?*
- *Did you have to change your point or points a lot or just a little to change your coefficient of determination?*
- *Did you make a graph to help you solve your problem? Can you think of an example of a set of two-variable data that would result in a graph like yours?*

✸ **BIG IDEA.** **There are a variety of ways to calculate a probability, both theoretical and experimental.**

> *Option 1:* Find the **odds** in favor of rolling doubles with a pair of dice.
>
> *Option 2:* If the odds in favor of an event are 1:3, find the odds in favor of the event happening twice in a row.

Students sometimes have difficulty distinguishing between odds and probabilities. Both options require students to find odds. *Option 2* requires understanding what compound events are and how to calculate probabilities associated with compound events.

No matter which option students choose, follow-up questions such as these would be appropriate:

- *How did you determine the total number of outcomes for your event?*
- *How was the probability of your event happening related to the odds in favor of the event happening?*
- *If you were to wager $10 on your event happening, how much money would you win?*

> *Option 1:* What is the probability of drawing two aces in a row from a deck of 52 playing cards?
>
> *Option 2:* What is the probability of drawing an ace from a deck of 52 playing cards, given that your first card was an ace?

Both options are **conditional probability** questions. Both options can be solved by using counting techniques or conditional probability formulas. One draws the condition to the student's attention more than the other.

Even though the tasks differ in complexity, there are a number of common questions that could be asked of all students. These include:

- *What information about a deck of cards did you need to know that was not given?*
- *Are your events independent or dependent events?*
- *What did you need to know to calculate the probability of your events happening?*
- *What strategy did you use to find the probability?*

Option 1: How would you set up a **simulation** to determine the probability of Will scoring three baskets in a row when his success rate is 80%?

Option 2: How would you set up a simulation to determine the probability of Will scoring three baskets in a row when his success rate is 50%?

Designing a simulation can help determine the probability of an event happening. This question allows students to consider a situation and how to set up a simulation with or without technology to make predictions on the probability of the outcome. **Option 1** requires a simulation that can create a split of 80% success and 20% failure, such as a spinner divided into two sections with one section having four times as much area as the other. Some students will find they can create simulations more easily with a 50% success rate, as in **Option 2**, because they can use a coin toss, with which they are more familiar.

Questions applicable to both tasks include:

- *How did you use the success rate in your simulation?*
- *How did you decide what device to use for your simulation experiment?*
- *How would you know how many times to repeat your experiment?*

Determine the number of different ways Kipper can go home from the park without backtracking.

Option 1:   **Option 2:**

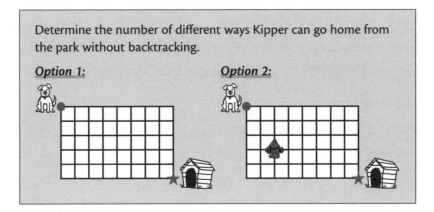

This parallel task provides a venue for using **counting principles** and knowledge about combinations to count paths or routes between two points. Both options have the same size grid, but **_Option 2_** has an added fire hydrant, through which the dog cannot move. This changes the number of possible routes, somewhat complicating the pattern students might use to help them solve **_Option 1_**.

Regardless of the option selected, follow-up questions such as these could be asked:

- *Which directions could you travel on the grid? Which would you travel?*
- *Is there always or not always more than one path to get from a point on the grid to another point?*
- *Are there always more ways to get to a point that is farther away than a point nearby?*
- *Describe the process you used to find the number of routes.*

> **_Option 1:_** A store has six types of doughnuts: plain, maple, frosted, chocolate, glazed, and jelly. How many different combinations of three doughnuts can you choose if more than one of your choices can have the same type of topping, but don't have to?
>
> **_Option 2:_** A store has five types of doughnuts: plain, maple, frosted, chocolate, and jelly. How many different combinations of three doughnuts can you choose if you cannot have the same topping twice?

The first option requires students to consider a broader range of possibilities. Both options are likely to elicit the use of counting principles to count combinations.

Relevant questions for both tasks include:

- *If you counted a combination maple, chocolate, plain, should you also count plain, chocolate, maple?*
- *How can you make sure you didn't forget a possibility you should include?*
- *What strategy did you use to find the total number of ways of choosing three doughnuts?*

✳ **BIG IDEA.** There are a variety of representations of a probability distribution.

> A dry foods company produces a 200 g bag of pasta. To try to ensure that each bag contains at least 200 g, the bags are produced so that the mean mass is 210 g, the standard deviation is 6 g, and the mass is distributed normally.
>
> *Option 1:* What is the probability that a bag selected at random has a mass of less than 200 g?
>
> *Option 2:* What is the probability that a bag selected at random has a mass between 190 g and 210 g?

Many statistical studies use data from a normal population. If a population is believed to be normally distributed, predictions can be made from the sample population. This parallel task requires students to use a standardized **normal distribution** table to determine probabilities that a bag will have a certain mass. *Option 2* is more complex because it asks students to determine the probability that a bag's mass is within a tighter range.

Even though the tasks differ, there are common questions that can be asked of all students. These include:

- *Will there be a lot of bags produced with a mass between 200 g and 210 g? Less than 200 g?*
- *Why can't you calculate an exact probability that the mass is 200 g?*
- *Did you use a graph to help you solve this problem?*
- *Should the dry foods producer adjust its manufacturing process?*

> 1,000 cookies are produced every hour, and 17% of the cookies get broken in the process and are sold for making pie crust.
>
> *Option 1:* Estimate the probability that in 1 hour, at least 18% of production will be sold as an ingredient for pie crust.
>
> *Option 2:* Estimate the probability that in 1 hour, exactly 205 cookies will be broken.

Both options of this parallel task require students to use a normal approximation for the discrete binomial distribution. *Option 1*, the smallest number of broken cookies to meet the stated criteria must be determined first. *Option 2* may be more difficult for students, because they must realize that finding the probability of exactly 205 broken cookies, using a normal distribution, means determining the probability of a value falling between 204.5 and 205.5.

No matter which option students choose, follow-up questions such as these would be appropriate:

- *Was a normal approximation valid for this binomial problem? How did you know?*
- *What are the mean and standard deviation?*
- *What did you need to know to carry out your estimation?*
- *Explain how you could use this estimate to determine a year's supply of broken cookies.*

SUMMING UP

MY OWN QUESTIONS AND TASKS

Lesson Goal: Grade Level: _____

Standard(s) Addressed:

Underlying Big Idea(s):

_____

Open Question(s):

_____

Parallel Tasks:

Option 1:

Option 2:

_____

Principles to Keep in Mind:

- All open questions must allow for correct responses at a variety of levels.
- Parallel tasks need to be created with variations that allow struggling students to be successful and proficient students to be challenged.
- Questions and tasks should be constructed in such a way that will allow all students to participate together in follow-up discussions.

The eight big ideas that underpin work in Data Analysis and Probability were explored in this chapter through 50 examples of open questions and parallel tasks, as well as variations of them. The instructional examples provided were designed to support differentiated instruction for students at different developmental levels, targeting two separate grade bands: grades 6–8 and grades 9–12.

The examples presented in this chapter only scratch the surface of possible questions and tasks that can be used to differentiate instruction in Data Analysis and Probability. Surely many new ideas have already come to mind. For example, students could further explore properties of normal curves or what information a provided box plot shape shows. A form such as the one shown here can serve as a convenient template for creating your own open questions and parallel tasks. The Appendix includes a full-size blank form and tips for using it to design customized teaching materials.

Conclusions

VIRTUALLY EVERY TEACHER of mathematics in the middle-school or high-school grades is faced with students who vary in mathematical readiness for what is being taught. The differences among students may be cognitive or may relate to learning style or preferences. Regardless of the source or nature of these differences, every teacher wants to help each student succeed in learning. One of the best approaches to fostering this success within today's classroom environment is through differentiating instruction. This book was written to help teachers accomplish this differentiation.

THE NEED FOR MANAGEABLE STRATEGIES

Realistically, teachers do not have the preparation time to develop or the instructional time to deliver alternative programs that specifically target the individual learning levels of the many students with whom they work. What is needed is a way for teachers to engage all students at an appropriate level with a single question or set of related tasks. This book has modeled two core strategies that could be all a teacher requires to meet the mathematical learning needs of most students. These strategies are open questions and parallel tasks.

Open Questions

A question is open when it is framed in such a way that a variety of responses or approaches are possible. To be useful in instruction, it is important that the question focus on an important mathematical concept or skill, that the teacher has considered the readiness of most of his or her students to deal with the question at some level, and that the teacher is comfortable accepting a broad range of appropriate responses from class members.

Parallel Tasks

Two or more tasks are parallel when they are designed to meet the needs of students at different development levels, yet address the same big idea and are close enough in context that they can be discussed simultaneously. In implementing a lesson with parallel tasks, it is important that the teacher be prepared with a series of at least two or three follow-up questions about student work that would be pertinent no matter which task a student had completed. Sometimes the teacher should decide which of the tasks each student is to complete, but at other times the students should be allowed to choose.

DEVELOPING OPEN QUESTIONS AND PARALLEL TASKS

The same general approaches can be applied to developing open questions and parallel tasks for mathematical instruction.

A Three-Step Process

Step 1: Review Examples. This book provides more than 270 open questions and parallel tasks that can serve as models for new questions and tasks. A review of the examples for a particular content strand at a particular grade band will give a good sense of appropriate instructional material. By using the examples, and the variations proposed, a teacher will have a firm foundation upon which to build.

Step 2: Create the Question or Tasks. The form provided in the Appendix can serve as a convenient template for creating new open questions and parallel tasks. Among other things, the form offers a reminder that coherent instruction is served through a focus on big ideas as a means of addressing content standards. Bearing in mind how students differ with respect to specific behaviors related to the targeted big idea, the teacher can create original questions and tasks or adapt them from an existing resource. The objective in all cases should be to produce materials that appropriately challenge a broad range of students, from those who are struggling to those who are "average" to those who are particularly capable.

Step 3: Plan for Follow-Up. The teacher must ensure that the conversation resulting from class work with an open question or set of parallel tasks is inclusive. Careful structuring of questions and coordination of tasks is crucial. In follow-up discussion, the teacher should try to build connections among the various levels of responses that students provide. Making these connections clear will help the struggling student see how the answer that he or she provided can be an important step toward a more sophisticated answer.

Creating an Open Question: An Example

Consider, for example, a situation in which the intent is to teach calculating the surface area of a cone in a classroom where students are still struggling with surface area of simple prisms or with areas of circles. Those struggling students might not be ready for the types of questions that were initially planned. To adapt the lesson for a differentiated approach, the teacher might choose a straightforward question from a text, such as:

Determine the surface area of a cone with a height of 8" and a radius of 4".

Recognizing that some students were not ready for this question, the teacher might ask, instead:

> *Choose three 3-D shapes that are 8" wide and 8" high. Determine their surface areas.*

The change is simple but important. Now every student can work at a level that benefits him or her. In the discussion of the problem as students share their work, every student in the class (no matter what shapes that person chose) gets to hear how other students handled the situation with the same or different shapes. A win-win!

Creating a Set of Parallel Tasks: An Example

Consider, again, the case of the surface area problem posed earlier. For some reason, a teacher might not be comfortable using the open question suggested above. In that case, it might be that the original problem involving the cone is posed, along with an alternate problem in which students calculate the surface area of a cylinder (or a rectangular prism). Students might be directed to a specific problem or offered a choice of which problem to complete. However, it is critical that the discussion of the problems be common to both groups. For example, the teacher could ask:

- *Did you need to know both the width and the height to calculate the surface area? Why?*
- *Could you calculate the surface area using one computation, or did you have to use a variety of computations? Why?*
- *How would the surface area have changed if the width had been doubled?*
- *Did you need to know what sort of shape it was to get the surface area? Why?*

There are two important points to note: (1) The context being the same is important to making a common discussion possible. (2) It is also important to set up the situation so that there could be common questions beyond simply *What did you do?*

Fail-Safe Strategies

Open questions can be created relatively easily by strategies such as giving an answer and asking for a question for which it is the answer, allowing students to choose numbers within a question, or asking how two items are alike and different.

Parallel tasks can be generated from a single original task by strategies such as changing the complexity of the numbers, shapes, graphs, patterns, equations, or measurements being employed or the complexity of the situations being addressed.

THE BENEFITS OF THESE STRATEGIES

Teachers may at first find it challenging to differentiate their teaching by incorporating open questions and parallel tasks. Before long, however, thinking about ways to open up instruction will become a habit when content that might be too difficult for some students must be taught. The techniques illustrated in this book will become valuable tools for adjusting instruction to accommodate students at all developmental levels.

In the beginning, it might make sense for a teacher to commit to using open questions at least once a week, ultimately aiming to use such questions every one or two class periods. And a teacher might commit to using parallel tasks at least once a week, working toward employing them even more frequently. Eventually, students might even be encouraged to create their own parallel tasks by changing the complexity of situations they are asked to address.

The use of these straightforward but powerful differentiation strategies will lead to a classroom with much higher engagement levels, much less frustration, and much more math learning, thinking, and talking—promoting a much more positive attitude toward math for most students.

Appendix:
Worksheet for Open Questions
and Parallel Tasks

STRATEGIES for developing open questions and parallel tasks are discussed in Chapter 1 and in the Conclusions. The form that appears on the next page can serve as a convenient template for creation of customized materials to support differentiation of instruction in math. The form is also available for download on the Teachers College Press website: www.tcpress.com.

The following fundamental principles should be kept in mind when developing new questions and tasks:

- All open questions must allow for correct responses at a variety of levels.
- Parallel tasks need to be created with variations that allow struggling students to be successful and proficient students to be challenged.
- Questions and tasks should be constructed in such a way that all students can participate together in follow-up discussions.

MY OWN QUESTIONS AND TASKS

Lesson Goal: **Grade Level:** _____

Standard(s) Addressed:

Underlying Big Idea(s):

Open Question(s):

Parallel Tasks:

Option 1:

Option 2:

Principles to Keep in Mind:

- All open questions must allow for correct responses at a variety of levels.
- Parallel tasks need to be created with variations that allow struggling students to be successful and proficient students to be challenged.
- Questions and tasks should be constructed in such a way that will allow all students to participate together in follow-up discussions.

Glossary

For each Glossary entry, the chapter and page of first occurrence of the term are given in brackets at the end of the definition.

AAS. A description that two triangles share identical measurements of two angles and the side adjacent to one of them but not between them; the triangles may or may not be congruent [Chap. 4, p. 115]

accuracy. The degree of closeness of a measurement to the true value [Chap. 5, p. 123]

acute angle. An angle less than 90° [Chap. 4, p. 95]

acute triangle. A triangle with three angles less than 90° [Chap. 4, p. 100]

algebra tiles. A set of concrete manipulatives designed to model the 2nd, 1st, and 0th powers of a variable; they can be used to model both positive and negative integer combinations of one or sometimes two variables [Chap. 2, p. 28]

algebraic expression. A combination of variables, operation signs, and numbers (e.g., $2x + 4$) [Chap. 2, p. 18]

algorithm. A standard procedure for performing a task (e.g., a standard procedure for adding numbers) [Chap. 3, p. 64]

amplitude. The distance from a sinusoidal function's mean value to either the minimum or maximum value of the function [Chap. 2, p. 37]

angle of elevation. The angle between the horizontal and the line from an object to the observer's eye [Chap. 5, p. 138]

apex. The top point of a cone, pyramid, or isosceles triangle [Chap. 4, p. 99]

arithmetic sequence. A sequence in which each term is a fixed amount greater than or less than the previous term [Chap. 2, p. 38]

arithmetic series. The sum of the terms of an arithmetic sequence [Chap. 2, p. 38]

ASA. A description that two triangles share identical measurements of two angles and the side between them; this results in the triangles being congruent [Chap. 4, p. 115]

asymptote. A line to which a curve gets arbitrarily close, but never reaches [Chap. 2, p. 54]

axis of a graph. A reference line drawn on a graph from which distances are measured; there is usually an *x*-axis, a *y*-axis, and sometimes a *z*-axis; plural is *axes* [Chap. 2, p. 34]

axis of symmetry. A reflection line in a two-dimensional figure; plural is *axes of symmetry* [Chap. 2, p. 31]

bar graph. A way to show and compare data by using horizontal or vertical bars [Chap. 6, p. 153]

base of a power. The symbol in a power expression that describes the factor that is multiplied by itself (e.g., 3 is the base in 3^4, indicating that 3 is multiplied 4 times) [Chap. 3, p. 86]

base of a shape. One of the two parallel polygons of a prism that are joined by parallelograms; the nontriangular face of a pyramid; the flat circular parts of a cone or cylinder [Chap. 4, p. 91]

base plan. A two-dimensional representation of a linking cube structure that shows the layout of the base of that structure and indicates the height of the structure at each part of the base [Chap. 4, p. 96]

base ten blocks. Blocks of different sizes representing place value columns; a larger block is 10 times the size of the next smaller block [Chap. 3, p. 81]

bearing. An angle from north that describes a direction; bearings go from 0° (north) to 360° [Chap. 4, p. 108]

bell curve. Often called a normal curve, this graph shows a typical distribution defined by its mean and standard deviation [Chap. 4, p. 103]

benchmark. A familiar measurement or number used for comparing other measurements or numbers (e.g., 1 inch, 1 foot, 25, 100, 500) [Chap. 3, p. 64]

bias. A distortion of results of a survey or experiment arising from inappropriate sampling or questioning techniques [Chap. 6, p. 155]

big idea. A statement that is fundamental to mathematical understanding; usually links many specific math outcomes [Chap. 1, p. 4]

binomial distribution. A discrete probability distribution of the number of successes of a given number of independent trials of the same two-outcome experiment with a given probability for each outcome [Chap. 6, p. 171]

bisect. Divide in half; an angle or a length might be bisected [Chap. 4, p. 102]

calipers. A measuring instrument with two hinged legs used to measure thickness or distance [Chap. 5, p. 126]

chord. A line segment joining two points on a circle [Chap. 4, p. 102]

circle graph. A graph divided into sectors, each proportionally representing a part of the data [Chap. 6, p. 153]

circle properties. A set of principles related to measurements of angles within a circle or lengths associated with chords of circles or tangents to circles [Chap. 4, p. 90]

circumference. The distance around a circle [Chap. 2, p. 57]

coefficient. The real number multiplier of a variable expression [Chap. 2, p. 22]

coefficient of determination. A measure of how well a particular model represents a set of data [Chap. 6, p. 181]

combination. A collection of items, usually of a given size, taken from a given set, for which the order of the items is irrelevant [Chap. 2, p. 64]

combinatorics. A branch of mathematics focused on the study of selecting, arranging, and counting subsets of objects selected from a given set [Chap. 6, p. 170]

common denominator. A single whole number that can be used to describe the denominators of equivalent fractions for all of the fractions in a set (e.g., 4 is a common denominator for $\frac{1}{2}$ and $\frac{3}{4}$, because $\frac{1}{2} = \frac{2}{4}$) [Chap. 3, p. 69]

common numerator. A single whole number that can be used to describe the numerators of equivalent fractions for all of the fractions in a set (e.g., 6 is a common numerator for $\frac{2}{3}$ and $\frac{3}{4}$, because $\frac{2}{3} = \frac{6}{9}$ and $\frac{3}{4} = \frac{6}{8}$) [Chap. 3, p. 69]

complementary angles. Angles with a sum of 90° [Chap. 5, p. 135]

complex conjugates. Two complex numbers that share the same real part and that have opposite imaginary parts (e.g., $2 + 3i$ and $2 - 3i$) [Chap. 2, p. 44]

complex numbers. A number of the form $a + bi$, where a and b are real numbers and $i = \sqrt{-1}$ [Chap. 3, p. 63]

composite number. A whole number with more than two factors (e.g., 6 is composite because it has factors of 1, 2, 3, and 6) [Chap. 3, p. 79]

compound event. An event whose probability of occurring can be determined by knowing the probabilities of separate events that make it up [Chap. 6, p. 154]

compound fraction. A fraction with a numerator or denominator that is in fraction or decimal form [Chap. 3, p. 69]

compress. Stretch with a fractional (less than 1) stretch factor [Chap. 2, p. 61]

concave shape. A shape in which some of the lines joining pairs of points inside the shape lie outside the shape; a shape with a "dent" [Chap. 4, p. 101]

conditional probability. A probability that an event will occur given that some other event or events have already occurred [Chap. 6, p. 182]

congruent. Being of the same size and shape [Chap. 2, p. 45]

conjecture. A hypothesis or guess that something is true [Chap. 4, p. 90]

connectivity. A property of a network that describes how nodes in the network are connected [Chap. 4, p. 90]

constant. A value in an algebraic expression that does not change even when the value of the variable changes[Chap. 3, 22]

coordinate grid. A grid made up of perpendicular lines; positions on the grid can be located by using two numbers in a plane—one describing the horizontal distance from a designated point and one describing the vertical distance from that point [Chap. 2, p. 32]

coordinates. Pairs of numbers on a plane or triples of numbers in space that describe the location of a point in a coordinate system [Chap. 2, p. 40]

cosine. A number describing the ratio of the length of the adjacent side to an angle to the length of the hypotenuse in a right triangle; abbreviated cos [Chap. 2, p. 43]

cotangent. The reciprocal of the tangent [Chap. 5, p. 133]

counting principles. Principles or strategies that allow one to determine the count of a large set without counting each member of the set individually [Chap. 6, p. 183]

cross-section. The shape created when a plane intersects a three-dimensional object [Chap. 4, p. 103]

cube root. The number that can be multiplied 3 times to result in a given number (e.g., the cube root of 64 is 4, because $64 = 4 \times 4 \times 4$) [Chap. 3, p. 84]

decompose. Cut a shape into smaller shapes [Chap. 4, p. 89]

dependent events. Events for which the outcome of one event is influenced by the outcome of the other [Chap. 6, p. 168]

derived measurements. Measurements in the SI system that are created by combining base measurements (e.g., speed is a derived measurement based on length and time measures) [Chap. 5, p. 124]

diagonal. A line segment connecting vertices in a polygon or polyhedron that are not already connected [Chap. 4, p. 93]

diameter. The widest distance across a circle [Chap. 2, p. 37]

differentiating instruction. Tailoring instruction to meet the needs of different types or levels of students in a classroom [Chap. 1, p. 1]

direct variation. A relation in which one variable is a multiple of the other [Chap. 2, p. 35]

discrete. Consisting of separate and distinct parts [Chap. 6, p. 180]

distance formula. A formula based on the Pythagorean theorem to determine the distance between two given coordinate pairs or triples [Chap. 4, p. 117]

domain. The set of all values for which the independent variable is defined [Chap. 2, p. 54]

dot product. A value calculated by multiplying corresponding elements of a vector and adding those values (e.g., the dot product for [3,4] and [−2,3] is 6, because $3 \times (-2) + 4 \times 3 = 6$) [Chap. 4, p. 118]

edge. A line segment formed where two faces of a three-dimensional shape meet [Chap. 4, p. 91]

ellipse. A shape formed when the total distance to two given points is fixed [Chap. 4, p. 103]

equation. A statement that two numbers or expressions are equal [Chap. 2, p. 17]

equilateral triangle. A triangle with three equal sides [Chap. 4, p. 105]

event. An outcome or set of outcomes in a probability experiment [Chap. 6, p. 162]

expected value. The sum of all possible outcome values, each multiplied by its probability of occurrence [Chap. 6, p. 171]

experimental probability. The probability that results from an experiment [Chap. 6, p. 154]

explicit formula. A description of the general term of a pattern in which the term value is related to the term's position in the pattern [Chap. 2, p. 30]

exponent. The symbol in a power expression that describes how many times a factor is multiplied (e.g., 4 is the exponent in 3^4, indicating that 3 is multiplied by itself 4 times) [Chap. 3, p. 63]

exponential function. A function in which an independent variable appears in an exponent [Chap. 2, p. 18]

exponent laws. A set of rules that can be used to simplify calculations involving exponents [Chap. 2, p. 42]

expression. A commonly used synonym for *algebraic expression* [Chap. 2, p. 17]

extrapolate. To predict a value by following a pattern beyond known values [Chap. 6, p. 165]

factor. One of the whole numbers that is multiplied to produce another number (e.g., 4 is a factor of 8, because $2 \times 4 = 8$); the process of breaking up a number or algebraic expression into factors [Chap. 2, p. 32]

factored form (of a function). A representation of a function as the product of terms in the form $(x - a)$ multiplied by a real number [Chap. 2, p. 45]

factor tree. A representation of the factors of a number that results in its prime factorization [Chap. 3, p. 79]

first differences. The difference between two consecutive y-values in a table of values in which the x-values are a fixed distance apart [Chap. 2, p. 31]

frequency distribution graph. A graph that divides raw data into intervals and shows the number of data elements or fraction of the data expected or observed in each interval [Chap. 6, p. 169]

function. A relation such that each element in the domain is associated with a single element in the range [Chap. 2, p. 17]

general term (of a pattern). A rule that describes how any particular element in a pattern is determined [Chap. 2, p. 19]

geometric sequence. A sequence in which each term is a fixed multiple of the previous term [Chap. 2, p. 58]

greatest common factor. The greatest value that is a factor of two or more given values [Chap. 3, p. 80]

hexagon. A shape with six straight sides [Chap. 4, p. 94]

histogram. A continuous bar graph; bars are joined at the sides because the data they represent are continuous [Chap. 6, p. 153]

hyperbola. A shape created when the difference between two given points is fixed [Chap. 4, p. 103]

hypotenuse. The longest side of a right triangle [Chap. 5, p. 127]

hypothesis. A conjecture [Chap. 6, p. 163]

image. The shape that results from reflecting, rotating, translating, or dilating an original shape [Chap. 4, p. 115]

improper fraction. A fraction with a numerator greater than or equal to its denominator [Chap. 3, p. 66]

independent events. A set of two or more events in which the outcome of one of the events is not influenced by the outcomes of the other events [Chap. 6, p. 168]

infinity. A concept describing an amount that is limitless [Chap. 3, p. 73]

integers. The set of counting numbers, their opposites on the other side of 0 on the number line, and 0 [Chap. 2, p. 19]

interpolate. To estimate a value between two known values [Chap. 6, p. 175]

inverse operation. An operation that undoes the work of another operation (e.g., subtraction is the inverse of addition, because subtracting 8 after adding 8 results in no change in a number) [Chap. 3, p. 64]

inverse sine. The angle whose sine is a given number [Chap. 5, p. 135]

irrational number. Any real number that cannot be expressed as the ratio of two integers [Chap. 3, p. 63]

irregular shape. A shape with sides that are not all equal in length [Chap. 4, p. 110]

isometric drawing. A two-dimensional representation of a three-dimensional shape in which equal lengths on the three-dimensional shape are drawn as equal lengths on the two-dimensional representation [Chap. 4, p. 97]

isosceles triangle. A triangle with two sides of equal length [Chap. 4, p. 92]

kite. A quadrilateral with two pairs of equal adjacent sides [Chap. 4, p. 101]

law of cosines. A relationship relating the side lengths of a triangle and the cosine of one of its angles [Chap. 5, p. 140]

law of sines. A relationship relating two side lengths of a triangle to measures associated with the opposite two angles [Chap. 5, p. 137]

least common multiple. The smallest value that is a multiple of all numbers in a given set [Chap. 3, p. 80]

leg. One of the two shorter sides of a right triangle [Chap. 5, p. 127]

limit. A value to which a set of values is drawing increasingly close [Chap. 4, p. 105]

line of best fit. A line that can be used to predict, reasonably well, the relationship between two variables [Chap. 6, p. 154]

linear. A relationship between two variables whose graph forms a line [Chap. 2, p. 18]

linear combination. A value obtained by adding multiples of given values (e.g., a linear combination of 3 and 5 is a number of the form $3x + 5y$, where x and y are integers) [Chap. 4, p. 106]

linear growing pattern. An arithmetic sequence [Chap. 2, p. 25]

locus. A description of the location of points that satisfy a given condition [Chap. 4, p. 117]

logarithm. The logarithm of a number to a given base is the power or exponent to which the base must be raised in order to produce the number (e.g., the base-10 logarithm of 1,000 is 3, because 3 is the power to which 10 must be raised to produce 1,000: $10^3 = 1{,}000$, so $\log_{10}1{,}000 = 3$) [Chap. 3, p. 85]

matrix. A set of data written in rows and columns; plural is *matrices* [Chap. 3, p. 87]

mean. A value for a set of numbers determined by adding them and dividing by the number of numbers in the set; average [Chap. 6, p. 153]

measurement error. The difference between a measured value and the true value of a measure [Chap. 5, p. 123]

median. A line joining a vertex of a triangle to the middle of the opposite side [Chap. 4, p. 116]; the middle number in an ordered set of numbers [Chap. 6, p. 153]

midpoint. A point that divides a line segment into two equal lengths [Chap. 4, p. 116]

mode. The most frequent value or values in a set of data [Chap. 6, p. 153]

multiple. A number that is the product of two other whole numbers (e.g., 6 is a multiple of 3) [Chap. 3, p. 79]

mutually exclusive events. Two or more events that cannot happen simultaneously [Chap. 6, p. 169]

negative integers. The opposite of the counting numbers 1, 2, 3, . . . on the other side of 0 on the number line (i.e., –1, –2, –3, . . .) [Chap. 3, p. 63]

net. A two-dimensional shape that can be folded to create a three-dimensional object [Chap. 4, p. 97]

network. A set of points along with the lines or arcs connecting them [Chap. 4, p. 90]

non-Euclidean geometry. A type of geometry in which the sum of the angles in a triangle is not 180° [Chap. 4, p. 90]

normal distribution. A theoretical frequency distribution represented by a bell curve [Chap. 6, p. 185]

obtuse angle. An angle between 90° and 180° [Chap. 5, p. 136]

obtuse triangle. A triangle with one obtuse angle [Chap. 4, p. 100]

octagon. A shape with eight straight sides [Chap. 4, p. 93]

odds. The ratio of the probability that an event will occur to the probability that it will not occur [Chap. 6, p. 182]

open questions. Broad-based questions with many appropriate responses or approaches; the questions serve as a vehicle for differentiating instruction if the range of possibilities allows students at different developmental levels to succeed even while responding differently [Chap. 1, p. 7]

order of magnitude. The relative size of a number in terms of powers of 10 [Chap. 3, p. 65]

order of turn symmetry. The number of times a shape can be turned and fit into its own outline by the time a full turn has been completed [Chap. 4, p. 91]

orthographic drawing. A view or set of views of an object from different perspectives, usually from the top, the right side, and the front [Chap. 4, p. 97]

outcome. A possible result for an experiment [Chap. 6, p. 154]

parabola. A shape created when the distance to a fixed point is set equal to the distance to a fixed line; a quadratic curve [Chap. 2, p. 32]

parallel lines. Two lines in a plane are parallel if they never meet; two nonvertical lines are parallel if and only if they have the same slope [Chap. 2, p. 42]

parallel tasks. Sets of two or more related tasks or activities that explore the same big idea but are designed to suit the needs of students at different developmental levels; the tasks are similar enough in context that they can be considered together in class discussion [Chap. 1, p. 7]

parallelogram. A quadrilateral with opposite sides that are parallel [Chap. 2, p. 53]

parameter. A constant in an equation that might be different in other equations of very similar form [Chap. 1, p. 33]

partial variation. A relation in which one variable is a constant amount greater than a fixed multiple of the other [Chap. 2, p. 35]

pattern rule. A way to describe how the terms of a pattern are defined [Chap. 2, p. 20]

pentagon. A five-sided polygon [Chap. 4, p. 110]

percentile. Values that divide a sample of data into 100 groups containing equal numbers of observations; for example, 60% of the data values fall below the 60th percentile [Chap. 6, p. 154]

perfect square. A number that is the product of a whole number multiplied by itself (e.g., 16 is a perfect square, because $16 = 4 \times 4$) [Chap. 3, p. 38]

perimeter. The distance around a shape [Chap. 2, p. 24]

period. The interval of the independent variable (often time) needed for a repeating action to complete one cycle [Chap. 2, p. 37]

periodic function. A function that repeats its values in regular intervals [Chap. 2, p. 41]

permutation. An ordered arrangement of a given set of items [Chap. 3, p. 64]

perpendicular lines. Lines that meet at a 90° angle [Chap. 2, p. 42]

phase shift. A horizontal shift of a sinusoidal function [Chap. 2, p. 38]

pictograph. A graph that uses symbols to represent quantities [Chap. 6, p. 158]

plane. A flat surface that extends infinitely [Chap. 4, p. 90]

polar coordinates. A pair of numbers describing the location of a point on a plane using a distance from a fixed point and the measure of an angle from a fixed line going through that point [Chap. 4, p. 107]

polygon. A closed shape with only straight sides [Chap. 4, p. 102]

polyhedron. A three-dimensional shape with only polygon faces; plural is *polyhedra* [Chap. 4, p. 90]

polynomial. An algebraic expression that is the sum of terms in which the variable is always raised to a whole number power [Chap. 2, p. 18]

population. All of the people or objects under consideration [Chap. 6, p. 159]

power. A number of the form y^x [Chap. 3, p. 64]

precision. The degree to which repeated measurements show the same result; normally if a smaller unit is used, the measurement is more precise [Chap. 5, p. 123]

prime number. A counting number with exactly two distinct factors [Chap. 3, p. 79]

prism. A three-dimensional shape with two identical polygon bases that are parallel and opposite each other and are connected by parallelograms [Chap. 4, p. 91]

probability distribution. A description of the probability of each possible outcome or outcome interval [Chap. 6, p. 154]

product. The result of a multiplication [Chap. 2, p. 29]

proportion. A description of how two numbers are related based on multiplication (e.g., the proportion 25:100 can also be described as the fraction $\frac{1}{4}$, the decimal 0.25, or the ratio 1:4) [Chap. 3, p. 63]

proportional reasoning. A focus on comparing numbers based on multiplication (e.g., thinking of 12 as four 3s rather than as, e.g., 10 + 2) [Chap. 3, p. 68]

pyramid. A shape with a polygon base and triangular faces that meet at a point [Chap. 4, p. 91]

Pythagorean theorem. A statement that the square of the length of the longest side of a right triangle is the sum of the squares of the lengths of the two other sides [Chap. 2, p. 58]

quadrant. One fourth of a coordinate grid; the first quadrant is to the right and above the center point; the second quadrant is to the left and above the center point; the third quadrant is to the left and below the center point; and the fourth quadrant is to the right and below the center point [Chap. 2, p. 41]

quadratic. A polynomial in which the highest power is 2 [Chap. 2, p. 18]

quadratic formula. A formula that can be used to determine the roots of a quadratic equation [Chap. 2, p. 32]

quadrilateral. A shape with four straight sides [Chap. 4, p. 93]

quartic equation. An equation involving a polynomial in which the highest power is 4 [Chap. 2, p. 44]

quartile. The value of the boundary at the 25th, 50th, and 75th percentiles of a frequency distribution that is divided into four equal groups [Chap. 6, p. 154]

quotient. The result of division (e.g., in the expression $8 \div 4 = 2$, the quotient is 2) [Chap. 2, p. 29]

radian. A real number that is the ratio of the distance around a circle to the radius of the circle [Chap. 5, p. 133]

radical. An expression written as a square root, cube root, and so forth [Chap. 3, p. 86]

radius. The distance from the center of a circle to any point on the circle [Chap. 2, p. 37]

range. The set of all values of the dependent variable [Chap. 2, p. 55]; the difference between the greatest and least values in a data set [Chap. 6, p. 160]

rate of change. The change in a dependent variable for a fixed change in the independent variable [Chap. 2, p. 52]

rational function. A function that can be written as the ratio of two polynomial functions [Chap. 2, p. 18]

rational number. A number that is the quotient of two integers [Chap. 3, p. 63]

rationalize. A process of multiplying the numerator and denominator of a fraction by an irrational number so that the denominator becomes an integer [Chap. 3, p. 86]

reciprocal. A fraction created by interchanging the numerator and denominator of a given fraction (or by dividing 1 by the given fraction) [Chap. 2, p. 32]

recursion formula. A pattern rule in which one term of a pattern is defined in terms of preceding terms [Chap. 2, p. 30]

reflect. To flip in a mirror line [Chap. 2, p. 46]

reflection symmetry. A type of symmetry created when half of a shape can be reflected onto the other half [Chap. 4, p. 91]

reflex angle. An angle between 180° and 360° [Chap. 5, p. 138]

regression coefficient. The slope of the line that most closely relates two correlated variables [Chap. 6, p. 154]

regular shape. A shape with all sides equal in length [Chap. 4, p. 94]

relation. A statement of how two variables are connected [Chap. 2, p. 17]

repeating decimal. A decimal characterized by the eventual repetition of a specific core of digits [Chap. 3, p. 68]

rhombus. A parallelogram with four sides of equal length [Chap. 4, p. 101]

right angle. A 90° angle [Chap. 4, p. 95]

right triangle. A triangle with a 90° angle [Chap. 2, p. 29]

right trapezoid. A trapezoid with a 90° angle [Chap. 4, p. 110]

rise. The increase or decrease in the y-values of a graph over a particular interval [Chap. 2, p. 52]

root. A number describing the base that must be repeatedly multiplied to form a given power (e.g., a square root or a cube root) [Chap. 3, p. 75]; the solution to an equation [Chap. 2, p. 32]

run. The increase or decrease in the *x*-values of a graph over a particular interval [Chap. 2, p. 52]

sample (population). A group of people or items selected from a population [Chap. 6, p. 163]

sample space. A listing of all possible outcomes of an experiment [Chap. 6, p. 172]

scalar multiplication. Multiplication by a single value [Chap. 3, p. 87]

scalene triangle. A triangle with three sides of different lengths [Chap. 4, p. 100]

scatterplot. A set of points plotted on a coordinate grid designed to make relationships between two variables evident [Chap. 6, p. 153]

scientific notation. A way of describing numbers as the product of a number between 1 and 10 and a power of 10 (e.g., 34.5 can be written as 3.45×10^1 in scientific notation) [Chap. 3, p. 63]

sequence. An ordered list of objects; often used as a synonym for *pattern* [Chap. 2, p. 30]

similarity. A relationship between two shapes such that one is the same shape but of a different size than the other; in effect, a relationship reflecting enlargement or reduction [Chap. 4, p. 89]

simple radical. A radical or multiple of a radical not combined with other radicals in a sum or difference [Chap. 3, p. 86]

simplify. To write as an equivalent expression that appears simpler (e.g., a fraction with no common terms, or an algebraic expression in which like terms have been collected) [Chap. 2, p. 28]

simulation. A model that can be used to mimic a real-life situation [Chap. 6, p. 183]

simultaneous linear equations. A system, or set, of linear equations, all of which must hold true simultaneously [Chap. 2, p. 29]

sine. A number describing the ratio of the length of the opposite side to an angle to the length of the hypotenuse in a right triangle; abbreviated sin [Chap. 2, p. 43]

sinusoidal function. A type of periodic function based on transformations of the equation $y = \sin x$ [Chap. 2, p. 37]

skewed data. A set of data that, when graphed, does not exhibit symmetry [Chap. 6, p. 164]

slant asymptote. An asymptote that is neither vertical nor horizontal [Chap. 2, p. 54]

slant height. A distance from any point on the base of a cone to the apex of the cone [Chap. 5, p. 140]

slope. The rise to run ratio of a line [Chap. 2, p. 29]

square root. A number that can be multiplied by itself to achieve a given number [Chap. 3, p. 75]

SSA. A description that two triangles share identical measurements of two sides and the angle adjacent to one of them but not between them; the triangles may or may not be congruent [Chap. 4, p. 115]

SSS. A description that two triangles share identical measurements of all three sides; this results in the triangles being congruent [Chap. 4, p. 115]

standard deviation. A particular measure of data spread based on deviations of data points from the mean [Chap. 6, p. 154]

standard form (of a quadratic). A quadratic written in the form $f(x) = ax^2 + bx + c$ [Chap. 2, p. 31]

stratified sampling. The process of dividing a population into subpopulations and selecting appropriate random samples from those subpopulations [Chap. 6, p. 163]

stretch. A transformation of a graph that widens or narrows a basic graph of a particular type [Chap. 2, p. 44]

substitute. To replace a variable with a given numerical value [Chap. 2, p. 23]

surface area. The total of the areas of the surfaces of a three-dimensional object [Chap. 4, p. 95]

symbol sense. A sense of the usefulness of algebraic symbolism and the ability to manipulate and make sense of symbols flexibly in many contexts [Chap. 2, p. 18]

symmetry. An attribute of a shape that indicates that the shape becomes exactly like itself if it is flipped or turned [Chap. 2, p. 32]

system of linear equations. A set of simultaneous linear equations [Chap. 4, p. 122]

table of values. A table that indicates values of an independent variable with corresponding values of the dependent variable [Chap. 2, p. 48]

tangent. The ratio of a the side opposite to an angle to the side adjacent to that angle in a right triangle [Chap. 5, p. 133]

tangram. A puzzle in the form of a square made up of seven specific pieces [Chap. 4, p. 94]

tessellation. A tiling of the plane with no gaps or overlaps [Chap. 4, p. 90]

tetrahedron. A pyramid with a triangular base; plural is *tetrahedra* [Chap. 4, p. 105]

theoretical probability. The probability that would be expected when analyzing all of the possible outcomes of a situation [Chap. 6, p. 154]

transformation. A motion that moves a shape from one position to another; translations, reflections, rotations, and dilatations are common motions studied [Chap. 2, p. 18]

translate. To move an object by sliding it horizontally and/or vertically [Chap. 2, p. 44]

trapezoid. A quadrilateral with one pair of parallel sides [Chap. 4, p. 101]

tree diagram. A graphic used to organize all combinations of events possible when more than one action is taken in an experiment [Chap. 6, p. 172]

trigonometric function. A function of an angle whose value is expressed as a ratio of two of the sides of the right triangle that contains the angle [Chap. 2, p. 18]

trigonometric ratio. A ratio of two side lengths in a right triangle associated with a given acute angle in that triangle [Chap. 5, p. 133]

trigonometry. A branch of mathematics based on the relationships between side lengths and angle measures in right triangles [Chap. 4, p. 89]

trinomial. A polynomial made up of three terms that are not like terms (e.g., $2x^2 - 6xy + 7$) [Chap. 2, p. 40]

turn symmetry. A type of symmetry created when a geometric shape can be turned about a point on a plane or a line in space to fit exactly its original outline in less than a full turn; also called rotational symmetry [Chap. 4, p. 91]

unit. A measurement used as the basis for describing other measurements [Chap. 4, p. 123]

unit analysis. Consideration of how the units of different measures combine to form new units (e.g., when multiplying a speed unit [distance/time] by a time unit, the result must be a distance unit) [Chap. 4, p. 124]

variable. A letter or symbol that represents an unspecified number in an equation and/or expression [Chap. 2, p. 17]

vector. A quantity specified by a magnitude and a direction; can be expressed in the form of a matrix with a single row or column [Chap. 4, p. 90]

vertex. The maximum or minimum point on a quadratic curve [Chap. 2, p. 31]; a corner of a shape [Chap. 4, p. 91]; plural is *vertices*

vertex form. A quadratic function in the form $y = a(x - h)^2 + k$, where the vertex is (h, k) [Chap. 2, p. 31]

vertical angles. Angles formed opposite to one another when two lines intersect [Chap. 4, p. 104]

view. A single perspective on a three-dimensional object or structure (e.g., a top view) [Chap. 4, p. 97]

volume. The amount of material required to construct an object [Chap. 3, p. 84]

***x*-intercept.** The first in a pair $(x,0)$ or triple $(x,0,0)$ of numbers that describes a point's horizontal distance from a reference point on a coordinate grid on a plane [Chap. 2, p. 32]

***y*-intercept.** The second in a pair $(0,y)$ or triple $(0,y,0)$ of numbers that describes a point's vertical distance from a reference point on a coordinate grid on a plane [Chap. 2, p. 29]

***z*-score.** A measure of the distance, in standard deviations, of a sample from the mean [Chap. 6, p. 154]

zero principle. A statement that $-1 + 1 = 0$ [Chap. 2, p. 40]

zone of proximal development. The range of potential learning that is beyond existing knowledge but that is accessible to a student with adult or peer support [Chap. 1, p. 2]

Bibliography

Becker, J. P., & Shimada, S. (1997). *The open-ended approach: A new proposal for teaching mathematics.* Reston, VA: National Council of Teachers of Mathematics.

Borgoli, G. M. (2008). Equity for English language learners in mathematics classrooms. *Teaching Children Mathematics, 15,* 185–191.

Bray, W. (2009). The power of choice. *Teaching Children Mathematics, 16,* 178–183.

Bremigan, E. G. (2003). Developing a meaningful understanding of the mean. *Mathematics Teaching in the Middle School, 9,* 22–27.

Bright, G. W., Brewer, W., McClain, K., & Mooney, E. S. (2003). *Navigating through data analysis in Grades 6–8.* Reston, VA: National Council of Teachers of Mathematics.

Charles, R. (2004). Big ideas and understandings as the foundation for elementary and middle school mathematics. *Journal of Mathematics Education Leadership, 7,* 9–24.

Ellis, M. W., & Malloy, C. E. (2008). *Mathematics for every student: Responding to diversity, Grades 6–8.* Reston, VA: National Council of Teachers of Mathematics.

Forman, E. A. (2003). A sociocultural approach to mathematics reform: Speaking, inscribing, and doing mathematics within communities of practice. In J. Kilpatrick, W. G. Martin, & D. Schifter (Eds.), *A research companion to Principles and Standards for School Mathematics* (pp. 333–352). Reston, VA: National Council of Teachers of Mathematics.

Friel, S., Rachlin, S., & Doyle, D. (2001). *Navigating through algebra, Grades 6–8.* Reston, VA: National Council of Teachers of Mathematics.

Gregory, G. H. (2005). *Differentiating instruction with style.* Thousand Oaks, CA: Corwin Press.

Gregory, G. H., & Chapman, C. (2007). *Differentiated instructional strategies: One size doesn't fit all* (2nd ed.). Thousand Oaks, CA: Corwin Press.

Gregory, G. H., & Chapman, C. (2008). *Activities for the differentiated classroom: Math, Grades 6–8.* Thousand Oaks, CA: Corwin Press.

Imm, K. L., Stylianou, D. A., & Chae, N. (2008). Student representations at the center: Promoting classroom equity. *Mathematics Teaching in the Middle School, 13,* 458–463.

Kabiri, M. S., & Smith, N. L. (2003). Turning traditional textbook problems into open-ended problems. *Mathematics Teaching in the Middle School, 9,* 186–192.

Karp, K., & Howell, P. (2004). Building student responsibility for learning in students with special needs. *Teaching Children Mathematics, 11,* 118–126.

Kersaint, G. (2007). The learning environment: Its influence on what is learned. In W. G. Martin, M. E. Strutchens, & P. C. Elliott (Eds.), *The learning of mathematics* (pp. 83–96). Reston, VA: National Council of Teachers of Mathematics.

Little, C. A., Hauser, S., & Corbishley, J. (2009). Constructing complexity for differentiated learning. *Mathematics Teaching in the Middle School, 15,* 34–42.

Lovin, L. A., Kyger, M., & Allsopp, D. (2004). Differentiation for special needs learners. *Teaching Children Mathematics, 11,* 158–167.

Mirra, A. (2009). *Focus in Grades 6–8: Teaching with curriculum focal points.* Reston, VA: National Council of Teachers of Mathematics.

Murray, M., & Jorgensen, J. (2007). *The differentiated math classroom: A guide for teachers, K–8*. Portsmouth, NH: Heinemann.

National Council of Teachers of Mathematics (NCTM). (2000). *Principles and standards for school mathematics*. Reston, VA: National Council of Teachers of Mathematics.

National Council of Teachers of Mathematics (NCTM). (2006). *Curriculum focal points*. Reston, VA: National Council of Teachers of Mathematics.

Pierce, R. L., & Adams, C. M. (2005). Using tiered lessons in mathematics. *Mathematics Teaching in Middle School, 11*, 144–149.

Pugalee, D. K., Arbaugh, F., Bay-Williams, J. M., Farrell, A., Mathews, S., & Royster, D. (2008). *Navigating through mathematical connections in Grades 6–8*. Reston, VA: National Council of Teachers of Mathematics.

Pugalee, D. K., Frykholm, J., Johnson, A., Slovin, H., Malloy, C., & Preston, R. (2002). *Navigating through geometry in Grades 6–8*. Reston, VA: National Council of Teachers of Mathematics.

Schifter, D., Bastable, V., & Russell, S. I. (1997). Attention to mathematical thinking: Teaching to the big ideas. In S. Friel & G. Bright (Eds.), *Reflecting on our work: NSF teacher enhancement in mathematics K–6* (pp. 255–262). Washington, DC: University Press of America.

Sheffield, L. J. (2003). *Extending the challenge in mathematics: Developing mathematical promise in K–8 students*. Thousand Oaks, CA: Corwin Press.

Small, M. (2005a). *PRIME: Professional resources and instruction for mathematics educators: Number and operations*. Toronto: Thomson Nelson.

Small, M. (2005b). *PRIME: Professional resources and instruction for mathematics educators: Patterns and Algebra*. Toronto: Thomson Nelson.

Small, M. (2006). *PRIME: Professional resources and instruction for mathematics educators: Data Management and Probability*. Toronto: Thomson Nelson.

Small, M. (2007). *PRIME: Professional resources and instruction for mathematics educators: Geometry*. Toronto: Thomson Nelson.

Small, M. (2009a). *Big ideas from Dr. Small, Grades 4–8*. Toronto: Nelson Education.

Small, M. (2009b). *Good questions: Great ways to differentiate mathematics instruction*. New York: Teachers College Press.

Small, M. (2010). *PRIME: Professional resources and instruction for mathematics educators: Measurement*. Toronto: Nelson Education.

Small, M., & Lin, A. (2010). *Big ideas from Dr. Small, Grades 9–12*. Toronto: Nelson Education.

Smith, M. S., Hughes, E. K., Engle, R. A., & Stein, M. K. (2009). Orchestrating discussions. *Mathematics Teaching in the Middle School, 14*, 548–556.

Tomlinson, C. A. (1999). *The differentiated classroom: Responding to the needs of all learners*. Alexandria, VA: Association for Supervision and Curriculum Development.

Tomlinson, C. A. (2001). *How to differentiate instruction in a mixed ability classroom* (2nd ed.). Alexandria, VA: Association for Supervision and Curriculum Development.

Tomlinson, C. A., & McTighe, J. (2006). *Integrating differentiated instruction and understanding by design*. Alexandria, VA: Association for Supervision and Curriculum Development.

Tomlinson, C. A., & Strickland, C. A. (2005). *Differentiation in practice: A resource guide for differentiating curriculum, Grades 9–12*. Alexandria, VA: Association for supervision and curriculum development.

Vygotsky, L. S. (1978). *Mind in society: The development of higher psychological processes*. Cambridge, MA: Harvard University Press.

Westphal, L. (2007). *Differentiating instruction with menus: Math*. Austin, TX: Prufrock Press.

Williams, L. (2008). Tiering and scaffolding: Two strategies for providing access to important mathematics. *Teaching Children Mathematics, 14*, 324–330.

Index

The Index is divided into two sections: the Index of Subjects and Cited Authors and the Index of Big Ideas. The Index of Subjects and Cited Authors covers instructional concepts as well as listings for content areas and grade bands; the names of authors cited in the text are also included. The Index of Big Ideas covers mathematical content broken down by big idea, providing access to the broad concepts presented. Individual mathematical terms are not indexed. The Glossary (pages 193–205) lists the primary mathematical terms featured in the text and instructional examples. Each Glossary entry ends with a chapter and page designator identifying the location of the first occurrence of each term.

INDEX OF SUBJECTS AND CITED AUTHORS

INDEX OF BIG IDEAS

How a shape can be composed and decomposed, or its relationship to other shapes, provides insight into the properties of the shape.

There are many representations of a geometric object or a relationship between geometric objects.

Locations of objects can be described in a variety of ways.

Measurement

A measurement is an explicit or implicit comparison.

The unit or tool chosen for a measurement can affect its numerical value as well as the precision of the measure.

The same object can be described using different measurements; sometimes the measurements are related, and other times they are independent.

Knowing the measurements of one shape can sometimes provide information about the measurements of another shape.

Measurement formulas allow us to rely on measurements that are simpler to access to calculate measurements that are more complex to access.

Data Analysis and Probability

To collect good data, it is necessary to decide what collection method is most suitable and how to best pose any questions required to collect the data.

Visual displays quickly reveal information about data.

Not only can information be read from visual displays, but conclusions can be drawn and inferences made.

How data are displayed can affect what conclusions are drawn from the data.

A summary statistic can meaningfully describe a set of data.

Sometimes it is reasonable to generalize from a sample of collected data.

There are a variety of ways to calculate a probability, both theoretical and experimental.

There are a variety of representations of a probability distribution.

About the Authors

MARIAN SMALL is the former Dean of Education at the University of New Brunswick. She has been a professor of mathematics education for many years and is a regular speaker on K–12 mathematics throughout Canada and the United States.

The focus of Dr. Small's work has been the development of curriculum and text materials for students and teachers of mathematics. She has been an author on seven text series at both the elementary and the secondary levels in Canada, the United States, Australia, and the country of Bhutan and a senior author on five of those series. She has served on the author team for the National Council of Teachers of Mathematics Navigation series, pre-K–2. For four years, she was the NCTM representative on the Mathcounts question writing committee for middle-school mathematics competitions throughout the United States. She is also a member of the editorial panel for the NCTM 2011 yearbook on motivation and disposition.

Dr. Small has written a text for university preservice teachers and practicing teachers, *Making Math Meaningful to Canadian Students: K–8*, as well as the professional resources volumes *Big Ideas from Dr. Small: Grades 4–8*, *Big Ideas from Dr. Small: Grades K–3*, and, most recently, *Big Ideas from Dr. Small: Grades 9–12*, the latter with Amy Lin, published by Nelson Education Ltd.

She has led the research resulting in the creation of maps describing student mathematical development in each of the five NCTM mathematical strands for the K–8 levels and has created the associated professional development program, PRIME. She has also developed materials and provided consultation focused on working with struggling learners and on teacher questioning in mathematics classrooms.

AMY LIN is a mathematics consultant for the Halton District School Board in Ontario, Canada. She supports K–12 teachers in mathematics instruction. Originally a research engineer, she became a high school teacher and was the recipient of the Ontario Teacher Award in 2004.

She has authored mathematics textbooks and student resources and has been a lead writer and developer for several Ontario Ministry of Education initiatives. Currently, she is working as a research lead for the Ministry of Education on a gap-closing project for struggling learners.